The Whats of a Scientific Life

Global Science Education

Series Editor
Professor Ali Eftekhari

Learning about scientific education systems in the global context is of utmost importance now for two reasons. Firstly, the academic community is now international. It is no longer limited to top universities, as the mobility of staff and students is very common even in remote places. Secondly, education systems need to continually evolve in order to cope with the market demand. Contrary to the past, when the pioneering countries were the most innovative ones, now emerging economies are more eager to push the boundaries of innovative education. Here, an overall picture of the whole field is provided. Moreover, the entire collection is indeed an encyclopaedia of science education and can be used as a resource for global education.

Series List:

The Whys of a Scientific Life
John R. Helliwell

Advancing Professional Development through CPE in Public Health
Ira Nurmala and Yashwant Pathak

A Spotlight on the History of Ancient Egyptian Medicine
Ibrahim M. Eltorai

Scientific Misconduct Training Workbook
John Gaetano D'Angelo

The Whats of a Scientific Life
John R. Helliwell

The Whats of a Scientific Life

John R. Helliwell

CRC Press is an imprint of the
Taylor & Francis Group, an informa business

CRC Press
Taylor & Francis Group
6000 Broken Sound Parkway NW, Suite 300
Boca Raton, FL 33487-2742

Printed on acid-free paper

International Standard Book Number-13: 978-0-367-23302-0 (Hardback)

Visit the Taylor & Francis Web site at
http://www.taylorandfrancis.com

and the CRC Press Web site at
http://www.crcpress.com

Science is a word derived from the

Latin word Scientia, *meaning 'to know'.*

Contents

Part III Examples of What Science Delivers or Will Deliver in the Future

Part IV Science and Mathematics: Across the Disciplines and Side by Side With Engineering

Part V Science Is a Process

Part VI A Trend: The Coming Together of the Sciences and the Social Sciences

Appendices My Reviews of Books Regarding the Whats of a Scientific Life

Preface

This book completes my scientific life trilogy of books on the *hows* (i.e. skills), the *whys* and the *whats* of a scientific life. How have I structured this third book? In Part I, I explore just what science is. To elucidate we must immediately consider the meaning of words like *knowledge*, *truth* and *certainty*. It is not surprising, then, that the subject of physics used to be known as 'natural philosophy'. As well as asking what physics is, I ask *what is chemistry?* and *what is biology?* My description of what science is, by its very nature of being done by people, includes considering *objectivity* and *ethics*.

In Part II, I discuss career situations in terms of types of obstacles faced, with my own examples.

In Part III, I describe examples of what science has achieved as well as some plans and opportunities about what will be achieved.

In Part IV, I look at the dependencies of science on mathematics, how scientific curiosity and the initiatives of funding agencies have spread across the disciplines and how science is placed side by side with engineering, leading to instrumentation, apparatus and computer software. These areas include my own career focus on, and roles within, crystallography and biophysics.

In Part V, I bring together thoughts about science as a process in avoiding or dealing with failures.

In Part VI, I describe the formation of the International Council for Science, a merged body of the International Council of Scientific Unions and the International Social Sciences Council. This joining of science and social science is a very significant step toward what will be world science.

In the appendices, I provide several reviews of books which impinge on the whats of a scientific life. One of these is a review of a book published in 1939, *The Social Function of Science*, by the polymath J. D. Bernal. I chose to write that review (see Appendix A1) because Bernal's book is still relevant today as a remarkably penetrating analysis of what science does, and what science could do, as well as a critique of how science was organised at that time.

In my two previous books [1, 2], I have documented my readings of various philosophy of science books, where I have found them practically useful in my scientific life. I honestly believe that my three books on the hows [1], the whys [2] and now the whats of a scientific life offer something different than a science research laboratory manual, a management tome or a philosophy of science treatise. My books offer an amalgamation of these.

Suffice it to say, I hope my three books are useful and interesting to a wide readership including my scientific colleagues and an interested public as well as schoolchildren.

REFERENCES

1. Helliwell, J. R. (2016). *Skills for a Scientific Life*. Boca Raton, FL: CRC Press.
2. Helliwell, J. R. (2018). *The Whys of a Scientific Life*. Boca Raton, FL: CRC Press.

Acknowledgements

As in my previous books I must thank all my students, staff and collaborators throughout my scientific research career for their contributions that have shaped my research and my career development. These include at the University of Manchester (1989–) and at the Daresbury Laboratory (1976–2008), leading research and education institutions to which I am especially grateful. I also thank the large number of agencies that have funded my research proposals. I am very grateful to the professional scientific organisations with which I have been involved throughout my career and notably the conferences that they organised many of which I assisted and a few of which I chaired. I have been especially involved in the International Union of Crystallography (IUCr) in various capacities, which has greatly diversified my experience, including representing IUCr at the International Council for Scientific and Technical Information and at the International Council for Science Committee on Data. These organisations have brought me into direct contact with scientists from a diverse range of science subjects and which I greatly appreciate.

I am grateful to various colleagues who have commented on my book manuscript: Katrine Bazeley, Member of the Royal College of Veterinary Surgeons; Emeritus Professor Carl Schwalbe of the Department of Pharmacy, Aston University, United Kingdom; Dr Anke Wilhelm of the Department of Chemistry, University of the Free State, Bloemfontein, South Africa; Dr Tom Koetzle, Chemistry Department, Brookhaven National Laboratory, Long Island, New York, United States; and Tom Lean of the British Library.

I thank Taylor & Francis for permission to reproduce in Appendices A1 and A5 my two book reviews that I had published in *Crystallography Reviews*. I also thank the International Union of Crystallography for permission to reproduce in Appendices A2, A3, A4 and A6 my book reviews that I had published in the *Journal of Applied Crystallography*.

About the Author

John R. Helliwell is Emeritus Professor of Chemistry at the University of Manchester. He was Director of Synchrotron Radiation Science at the Council for the Central Laboratories of the Research Councils (CCLRC). Professor Helliwell has served as President of the European Crystallographic Association (ECA). He was awarded a DSc degree in physics from the University of York in 1996 and a DPhil in molecular biophysics from the University of Oxford in 1978. He is a Fellow of the Institute of Physics, the Royal Society of Chemistry, the Royal Society of Biology, and the American Crystallographic Association. He is an Emeritus Member of the Biochemical Society, and in 1997, he was made an Honorary Member of the National Institute of Chemistry, Slovenia. In 2000 he was awarded the Professor K Banerjee Centennial Silver Medal by the Indian Association for the Cultivation of Science in Calcutta, India. In 2015 he was elected a corresponding member of the Royal Academy of Sciences and Arts of Barcelona, Spain. He was made an Honorary Member of the British Biophysical Society in 2017. That same year, he became a Faculty 1000 Member, charged with highlighting significant science publications. He was a Lonsdale Lecturer of the British Crystallographic Association in 2011, the Patterson Prize Awardee of the American Crystallographic Association in 2014 and the Max Perutz Prize Awardee of the European Crystallographic Association in 2015. In 2019, he was elected an Honorary Life Member of the British Crystallographic Association. Professor Helliwell has published more than 200 research publications and two research monographs. He has held leading roles within the International Union of Crystallography, most recently as Chairman of its Committee on Data and Chairman of its Book Series. He has chaired various science advisory committees at synchrotron X-ray and neutron facilities around the world including in France, Spain, Japan, the United States, Australia and Sweden.

Part I

Introduction

1

What Is the Scientific Life?

1.1 Scientific Objectivity, Truth and Certainty

The place to start is to query *what is science and what does it do*? This would then lead to answering the question of what a scientific life is. As a working definition, *science* is a word derived from the Latin word *scientia*, meaning 'knowledge'. At my local library in Horbury, West Yorkshire, when I was a boy, the entrance door had above it "Knowledge is power", a statement which always fascinated me. Power over what? I asked myself. That no mention was made specifically about science and the link to the Latin word *scientia* and to knowledge did not register with me then. Science finds out knowledge in the form of testable explanations and predictions about the universe. Since it is done by people, I have found it fascinating to consider whether science can be objective. I think that through measured data, and its careful preservation, it can be.

Whilst we are asking *What is science?* I think it interesting to consider *What is art?* too. I am not at all qualified to answer this question. I saw, though, that *The Times* on March 28th, 2019 quoted the world-renowned artist Pablo Picasso as follows: "We all know that Art is not truth. Art is a lie that makes us realise truth". If science is by definition knowledge, the matter raised by Picasso is, in effect, can science reach truth? My answer is that I don't think that science can realise truth, even though it reveals knowledge and can in the way I suggest through data preservation be objective. Quantum mechanics opens a new window on truth in that it includes uncertainty, by which I mean it is not possible to know with certainty the position and momentum of a particle simultaneously. This is Heisenberg's uncertainty principle [1]. So, if one cannot know those simultaneously, science can never reach truth.

Of course, one can play with the words and say that *the truth is uncertainty*. What should we also say about certainty or uncertainty in the physical world? In science we make experimental measurements. We seek to do the best job possible in our experimental design both in removing extraneous effects and taking as many repeat measurements as possible. The former is an attempt to remove any systematic error sources and the latter is an attempt by averaging the measurements to minimise random errors. Thus, with one experimental method a scientist reaches as precise an estimate as possible. But is that experimental estimate from one method *accurate*? Physicists make a distinction between precision and accuracy. Accuracy can be investigated by using a different experimental method to check the first experimental method's result. Ultimately, even with more than one method, there are errors at some level always remaining. Again we see that certainty is not achievable. The truth is that the physical world is uncertain. Can science be objective? The answer, I think, is that with modern day raw data archiving objectivity is possible. Also, we must ask: do scientists communicate their results objectively? I illustrate the process of communicating science in scientific articles and the role of data underpinning an article in Figure 1.1. With accompanying data, readers do not have to only accept the authors' words describing a result, but if needs be can undertake their own calculations and offer a different interpretation of those data. Does having the raw data available provide complete objectivity? This is a long-debated question (e.g. see [2]). Even where primary data being well measured, it is argued, an investigator is required to make the apparatus well calibrated and thereby the data can still be considered to that degree subjective!

Let me conclude this section by highlighting what science achieves. In spite of the uncertainties that I describe above, and its subjectivities, science

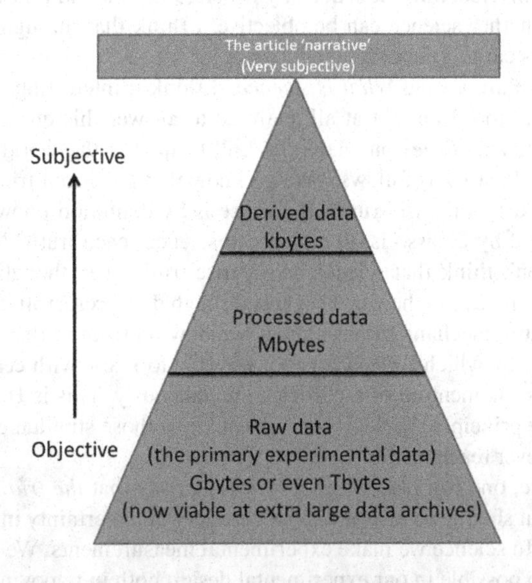

FIGURE 1.1 The role of the article in communicating science and achieving scientific objectivity through archiving the raw, i.e. primary, experimental measurement data sets underpinning it.

achieves firm facts such that one can rely on its applications. Whether one considers medicines, computers, smartphones, chemicals, food production, energy supplies, etc., science works.

1.2 Ethics in Science

Having considered what scientific objectivity is as well as what is possible regarding certainty and truth, let us now proceed to another big and important word to consider, *ethics*. Scientists in going about their daily work in the laboratory are, or should be, guided by ethics. Medical doctors have a formalised Hippocratic Oath [3]. Scientists do not have such a formalised oath, although establishing one and discussions of what it would contain have been considered [4]. In my two previous books on the scientific life I addressed how we do things [5] and why we do the things that we do [6]. In my *Hows* book I had an extensive chapter on ethics [7], which described both the ethics in undertaking our work, including our relationships with others, and, secondly, the ethical implications of some of the discoveries that science and scientists make.

The absence of a Hippocratic Oath for scientists is exceedingly strange to me. Research-based institutions, be they universities or research institutes, do have formal policies of what constitutes research malpractice; the University of Manchester's can be found in reference [8]. I was an elected Member of the Senate of the University representing the Faculty of Science and Engineering when this policy was approved, including by me.

It is obvious that a scientist should do no harm, to harness a term from the medical doctor's Hippocratic Oath. But what should one do to optimise the good that one does? Science as a career is in the end a calling. There can be wealth earned from one's discoveries, and fame in a few cases, but in general the scientific life is a devotion to a calling that there is possibly a greater good. Max Perutz wrote a book entitled *Is Science Necessary?* [9]. He documented his answer of a very firm, *Yes it is necessary*, with examples drawn from the areas of *Science and Food Production*, *Science and Health* and *Science and Energy*. He explained the greatly extended lifespan now available to people through science, the greatly improved comfort and control over nature, and greater economic wealth.

Maybe, to 'do no harm' the purest course of action is, like safety, not to venture forth and do anything. Likewise for the employer, as in the BBC's *Yes Minister* comedy programme episode which featured the story about the hospital with no patients but perfect metrics: no complaints, no inadvertent injuries and no fatalities. But there is one's scientific curiosity to satisfy, research to be done and discoveries to be made applying the training one has received in the lecture halls and the teaching laboratories. As on a sunny day, to venture forth is exciting. There are in any case good guidelines for avoiding pitfalls. Local rules govern how one should operate one's equipment and apparatus. Laboratory notebooks are to be kept up to date; the time-honoured way involves a top sheet and a duplicate underneath 'carbon copy' so researcher and laboratory director can retain a copy for future publication or the patent application or both in a strict time sequence, patent being before publication. In such ways the daily process of work and discovery in the laboratory, or at the computer 'calculations engine', or deriving your equations are what you are about.

But to say "it is obvious that a scientist should do no harm" is much easier to interpret for a medical doctor than for a scientist. Clearly if a patient gets better without side-effects following treatment, then the doctor has successfully done no harm. In science it can be much more complicated, with 'good' discoveries leading to unforeseen and sometimes bad consequences that damage the environment, the planet, individuals, etc. In this book, I examine the question of how science strives to avoid failures in various places and bring these together in one section in the final chapter.

1.3 Some Practicalities Day-to-Day, Week-to-Week and Month-to-Month

In the scientific researcher's workplace, the *what to do now* for funded research is heavily guided by the research grant proposal that has passed the funding agency's peer review approval and final committee procedures. The employer has accepted the research grant with which their principal investigator is eager to move forward. Modern research grant proposals have detailed timelines of objectives and milestones. It is likely that they are achievable as the funding agency schemes and their committees are usually conservative with respect to committing public money or the charitable foundation's assets. The issues are more about hiring properly skilled staff for the work envisaged. So, all should be well. The majority of proposals are however rejected, even though there are very few poor proposals. The research grant success rates are typically 25% or as low as 10%. Proposers, in my experience, myself included, do not just give up on those areas of unfunded work; they comprise cherished ideas, a detailed articulation and maybe also actual feasibility studies. So the *what to do now* includes those unfunded areas too. Their scope must be reduced and the length of time needed to take them on extended, and most importantly proposers must figure out how to fit them in with their daily/weekly/monthly/yearly *what to do lists*.

One of the biggest preoccupations is *what to do next*. The staff and PhD students, and the final year undergraduate project students, will focus daily on their next step career choices. Questions often arise such as *Stay in research? Move to another laboratory? Change research fields? Change country?* For final year undergraduate students, the top-rank ones who could go into scientific research wonder if they should. One of my best ever tutorial groups (chemistry with studies in North America) had everyone able to consider doing a PhD as their next step. One of this group, who was female, said quite plainly: "But look around you Prof, there are so few women scientists. It makes no sense as a career step for women". I was at that time also the department's Gender Equality Champion and Chairman of our Athena SWAN [10] working group. I honestly replied, "We know it is poor, scandalous actually, compared with the percentage of female chemistry graduates we train (44%) but led by Athena SWAN we are working hard to change the current situation". Evidently it was too much a case of 'jam tomorrow' for her. It was tragic I felt. So, part of what we do in our schedules is not only to reflect on our scientific projects but also to analyse what is right and what is wrong in the scientific environment we are in and make time to effect change where needed.

Diversity extends not only to gender. It also extends to Black, Asian and other Minority Ethnic groups (BAME), who may well be under-represented in your workplace. I was glad to secure a PhD studentship for an Asian student who had done an excellent final year project with me and my wife and colleague, Dr Madeleine Helliwell. In the end his father refused to let him accept it because he should get a 'proper job'. With the percentage of our PhDs ending up with permanent jobs in science being very low the father had a point. The funding agencies are trying to effect positive change here. An analysis of the biomedical research environment and lack of a career progression of all PhDs or postdocs in the United States is very relevant to this [11]. This in-depth study and commentary documents the undue proliferation of the number of PhDs and postdocs in biomedical research in the United States and then goes on to make specific recommendations for improvements. These include the need for a predictable budget for a funding agency, such as a five-year commitment from the U.S. government (rather than an annual approval method); reducing the number of trainee scientists; increasing the salaries of postdocs; and encouraging the briefing of such scientists about alternative careers. Many of these are already covered in the British funding environment. An additional aspect of this report is the importance of recognizing that the principal investigator (PI) scientists are not just writing machines of grant proposals and of publications, but should be encouraged/required to continue their own research and the regular updating of their research skills. Indeed, I firmly encouraged such an approach in my *Skills for a Scientific Life* book [5], not least as the poor grant proposal success rates obviously commend that one does so or else abandon various pieces of one's carefully planned research. Overall, most importantly, whilst we must present our results as well as possible and, yes, we must strive to win funding with the best written proposals, the heart of the skills for a scientific life are our own practical engagement, or otherwise we become 'solely' a research manager rather than a research scientist.

I hope that like my previous books, scientific monograph or careers books, you find this new book useful. I feel I have been lucky in my career but have had 'tight corners' to negotiate my way out of too. If you wish to listen to an audio account of my life in science, question-and-answer style, this is available at the British Library, recorded in 2017 [12]. (The interviewer was Tom Lean of the British Library.)

REFERENCES

1. Heisenberg, W. (1927) "Über den anschaulichen Inhalt der quantentheoretischen Kinematik und Mechanik" Zeitschrift für Physik 43 (3–4): 172–198. (English translation of the title: About the descriptive content of quantum theoretical kinematics and mechanics).
2. See section 2.3 of https://plato.stanford.edu/entries/scientific-objectivity/.
3. https://en.wikipedia.org/wiki/Hippocratic_Oath Accessed January 17th 2019.
4. King, David (2007) "A Hippocratic oath for scientists" https://en.wikipedia.org/wiki/Hippocratic_Oath_for_scientists Accessed January 17th 2019.
5. Helliwell, J. R. (2016) *Skills for a Scientific Life*. Boca Raton, FL: CRC Press.
6. Helliwell, J. R. (2018) *The Whys of a Scientific Life*. Boca Raton, FL: CRC Press.
7. Helliwell, J. R. (2016) Chapter 31 in *Skills for a Scientific Life*. Boca Raton, FL: CRC Press.
8. "The University of Manchester policy on research practice" http://documents.manchester.ac.uk/display.aspx?DocID=611 Accessed January 17th 2019.
9. Perutz, Max (1991) *Is Science Necessary? Essays on Science and Scientists*. Oxford: Oxford University Press.
10. https://www.ecu.ac.uk/equality-charters/athena-swan/ Accessed January 17th 2019.
11. Alberts, B., Kirschner, M. W., Tilghman, S. and Varmus, H. (2014) "Rescuing US biomedical research from its systemic flaws" *Proceedings of the National Academy of Sciences* 111 (16): 5773–5777.
12. Helliwell, John "(13 episodes) in an oral history of British science" https://sounds.bl.uk/Oral-history/Science/021M-C1379X0122XX-0001V0 Accessed January 17th 2019.

2

What Is Physics?

We can consider the founding father of physics to be Sir Isaac Newton (1642–1726). Newton's physics became known as 'classical physics' and reigned until Einstein questioned the basis of understanding in physics by asking a key question like, "What if the speed of light is finite?" In another challenge to Newtonian classical physics, Einstein's general theory of relativity led to explaining the bending of light in a gravitational field. This was first observed in a total solar eclipse on May 29th, 1919 by a team led by Sir Arthur Eddington. These different areas of physics marked a transition from a Newtonian understanding of the universe to an Einsteinian one. Newtonian mechanics was an approximation still valid at speeds much less than the speed of light, which remains a defining boundary between the two.

Isaac Newton's book *Philosophiae Naturalis Principia Mathematica* (1687) [1] translates into English from Latin as *Mathematical Principles of Natural Philosophy*. 'Natural philosophy' then is a term commonly used before the fields of science like physics, chemistry and biology were separately distinguished. To couple the word 'philosophy' with 'science' in this term 'natural philosophy' is a very interesting combination. It links with the definition that I introduced earlier that *science is a word derived from the Latin word scientia, meaning 'to know'*. That physics is at the heart of this definition can be illustrated in the following way.

During summer vacation in 1971, having just finished my school years in July and waiting to start my physics degree at York University in October, I started in on the summer reading matter prescribed by the York Department of Physics. This included George Gamow's book *Mr Tompkins in Paperback* [2]. It explains the central concepts in modern physics, from atomic structure to relativity, and quantum theory to fusion and fission, but in a most imaginative way. One of the ideas where his imagination is in full play is when he invites the reader to consider their everyday world but where Planck's constant is not the incredibly small value of 6.626×10^{-34} Js but rather is unity. (Planck's constant, h, features in the equation relating the energy of a photon, E, to the frequency of light, ν, namely $E=h\nu$.) Playing billiards in such a world with h=1 Js thereby meant no certainty with one's shots. It illustrated for me in a superb way the meaning of Heisenberg's uncertainty relation that one cannot know exactly at the same time both the position and the momentum of a particle, or a billiard ball in a Planck-constant-of-1 type of world.

Obviously the ideas in George Gamow's book [2] formed a pivotal transition for me from my school physics to my university physics course. How had I been prepared at school? Our set book [3] was *Advanced Level Physics* by Nelkon and Parker. It had more than 1,100 pages. Each chapter had numerical exercises to test our understanding and further learn from. Our lessons followed a syllabus set by the Northern Universities of England Joint Matriculation Board and were underpinned by this book and weekly practical physics laboratory classes. I really enjoyed my lessons, my laboratory experiments and the book by Nelkon and Parker. I had complete confidence in our physics teacher, Mr Bear. My preparation for the A level standard, which I did at ages 16 and 17, was the 'ordinary level' physics course (known as O level) when I was 14 and 15 years old. Our O level physics teacher, Mr Flint, was an enthusiast for the NASA Apollo programme. This enthused his pupils, myself included. But he did not complete the O level physics syllabus, leaving out completely electricity and magnetism. My grade at O level suffered, achieving a middle rank 'grade 3'. My A level grade was a 'B', i.e. second rank. I felt I was catching up, in effect. My university physics honours degree classification in July 1974 was 'first class'. My university physics course in the third year split into a choice of either the experimental physics option or the theoretical physics option; I took the experimental option. We selected experimental teaching laboratory projects each week and, being a small class of 32 students, I undertook each week's laboratory project on my own and wrote it up on my own as well, and it was then marked; I have kept all my laboratory notebooks. The York University Physics Department Head was Professor Oliver Heavens. He wrote to my school to say how I had performed in my final degree examinations and also encouraged me to join the Institute of Physics as a graduate member, which I gladly did. I really appreciated those actions of Professor Heavens. Later in my scientific career I applied for and was admitted as a Fellow of the Institute of Physics. I now serve the Institute of Physics regularly on its panels, judging individual applications for Fellowship.

Who do I regard as the greatest physicist? I am fascinated by Albert Einstein and his contributions to physics. I especially like the biography by Abraham Pais [4], which includes chapters describing in detail particular physics topics addressed by Einstein. One of these topics which I had not expected was Einstein's estimate of Avogadro's constant based on Brownian motion of molecules in a solution, and its comparison with measurements of the viscosity of a solution of sucrose; see Section 2 of reference [4].

REFERENCES

1. Newton, Sir Isaac (1687) *Philosophiae Naturalis Principia Mathematica.* Cambridge: Cambridge University Press.
2. Gamow, George (1965) Chapter 7 in *Quantum Billiards of Mr Tompkins in Paperback.* Cambridge: Cambridge University Press.
3. Nelkon, Michael and Parker, Phillip (1995) *Advanced Level Physics.* The book is now in its 7th edition. Portsmouth: Heinemann.
4. Pais, Abraham (2005) *Subtle Is the Lord: The Science and the Life of Albert Einstein.* Oxford: Oxford University Press.

3

What Is Chemistry?

At age 15, as is the tradition in England, I approached the deadline at my school for having to choose between the arts and humanities versus the sciences for my studies in my final two years before university. To assist pupils in making this clearly important choice, my school held a teachers–parents evening. It was a special evening because it was a chance for parents and children together to meet the school's careers adviser. By this time, we had already had school visits to learn about our job prospects, including going down the local coal mine and seeing the production line at the local shirt factory. These visits didn't match my hopes for continuing my learning. The school's careers adviser, Mr Taylor, had been the school's chemistry teacher but changed roles late in his teaching career. He asked me about my plans. I said I was thinking of going further with my studies and choosing history, geography and mathematics, which were my best subjects in terms of exam grades thus far. He said that it would be better if I included chemistry in my choices as chemistry was better for job opportunities. So, in this context, 'What is chemistry?' can be described as the subject with the best job opportunities! In the end I did select chemistry, along with mathematics and physics, to study in my last two years of school.

For these last two years our chemistry teacher was Dr Taylor, no relation to Mr Taylor. He held the attention of all of us by talking quietly and walking to and from the blackboard exceedingly slowly. It was all quite soothing. He was also an inspiration by bringing home to us the systematic way in which the periodic table organised all our thoughts on the chemistry of the elements, their properties and relationships being neatly ordered in rows and columns. To me, at the time and continuing now, the periodic table is one of the main achievements of all of science and of humanity. Indeed, as I write this, 2019 is the United Nations International Year of the Periodic Table, and rightly so. I was awarded the senior chemistry prize of the school in June 1971. It must have been disappointing to Dr Taylor that I applied to university to study physics rather than chemistry. Indeed, I spent the book token that came with my chemistry prize on a book [1] entitled *Atomic and Nuclear Physics*. But, today, Dr Taylor would be pleased, I think, to know that the final stage of my scientific career is as a professor of chemistry, now Emeritus, at the University of Manchester.

I can't remember what my school chemistry set books were, unfortunately. I do recall that, unlike my school physics studies, where we had one book (Nelkon and Parker, which I described in my 'What is physics?' chapter above), we had separate books for inorganic, organic and physical chemistry. When I became a chemistry academic I discovered the book *General Chemistry* by Linus Pauling [2]. So, the foundations of all of chemistry could be brought together in one volume! More than that, Pauling included biochemistry and also the myriad ways in which the fundamental particles inside the nucleus of the atom come together. Pauling obviously thought of such interactions in particle physics as still, quite simply, chemistry!

School laboratory experiments back then were more adventurous than they are today, it now being a much more chemical-safety-conscious era. In my role as a chemistry academic, I interviewed many schoolchildren who had applied to enter our degree course. I would ask them how their studies were going and what laboratory experiments they were doing. They would always answer: titrations and watching for colour changes.

Transitioning from being an academic physicist, often jointly with my role of developing instrumentation and methods at the United Kingdom's Synchrotron Radiation Source, to being a chemistry professor at the University of Manchester I found to be not so easy. Unlike my physics departments at Keele University and York University, where I taught and tutored courses and topics in my area of expertise (such as physical optics, biophysics, X-ray crystallography, Nuclear Magnetic Resonance [NMR] spectroscopy, mathematical methods and Fortran computer programming), I tutored all of inorganic chemistry. My lecture courses at Manchester were in my areas of crystallography or structural chemistry. But in both my academic physics and chemistry roles, my supervisions of laboratory experiments were across a diverse range of practicals. Since I was a chemistry professor I decided I should apply to become a member of the Royal Society of Chemistry (RSC) and I was duly admitted as such. A colleague of mine, Dr David Machin, said to me that since I was a chemistry professor, I should have applied to enter as a Fellow. I waited a couple of years and applied and was admitted as a Fellow of the RSC.

REFERENCES

1. Littlefield, T.A. and Thorley, N. (1968) *Atomic and Nuclear Physics*. New York: van Nostrand.
2. Pauling, Linus (2003) *General Chemistry*, now in a 3rd edition. New York: Dover Publications.

4

What Is Biology?

Ottoline Leyser (2019) recently stated, "The defining feature of biology during the past few decades has been figuring out details of the parts. But biological systems don't think they have parts" (quoted in [1]).

At age 13, as I approached a biology lesson, I was apprehensive. In the previous lesson our teacher had said to my class, "Next lesson you will each be able to dissect an earthworm". But I really didn't want to do that. I enjoyed gardening at home, planting seeds such as broad beans and seeing the plant shoots emerge from the ground and then grow to fully sized plants giving us lovely fresh broad beans to eat. So, why were we, in our biology laboratory class, having to kill an earthworm? At that time, as I mentioned, I was 13 years old. So, when at age 14 I had to select my O level subjects, eight in all, I decided I would drop biology. When I started my DPhil at Oxford University in October 1974 on the X-ray crystallography of the sheep liver 6-phosphogluconate dehydrogenase enzyme, I returned to biology as a subject, albeit at this stage biochemistry. I had decided to do a DPhil involving protein crystallography because I had enjoyed very much the lectures given by Dr Peter Main in the biophysics option course of my York University physics degree course. During this time I came across the book by the famous physicist Erwin Schrodinger [2] entitled *What Is Life?* I had been immediately intrigued because his name was associated with the famous equation of quantum mechanics, 'the Schrodinger equation'. I regarded it as a defining book for bridging physics and biology.

Later in my career as a scientist I was approached by Dr Peter Zagalsky of Royal Holloway College, University of London, for a collaboration on the X-ray crystal structure analysis of the blue–black coloration protein crustacyanin. Peter had purified the proteins from the lobster shell. Of course, the lobster had to be killed to provide its shell for the purification. But I didn't see that step in the research study. Nevertheless, it did seem to me that not being in on the killing of the lobster was in itself not enough to make it acceptable. I did consult a specialist on the question and he argued that the lobster does not 'feel' pain. Again I was unconvinced by this statement because how would he 'know', specialist or not? Ultimately one comes to the conclusion that the lobster's death is to enable fundamental research, and if it's done humanely then perhaps that is acceptable. Also, of course, members of the public endorse a restaurant boiling lobsters alive so as to eat them. At which reminder it makes one wish to become a vegetarian.

The proteins in the crustacyanin were crystallised by our collaborator, Professor Naomi Chayen. When we determined the crystal structure and published it in the journal *PNAS* [3], we also issued a press release. There was a lot of interest in our results because they explained why a lobster changes colour upon being cooked. We were contacted by many people who read about our research in newspapers or heard about it on their radios. These contacts included reports about fishermen occasionally catching lobsters with strikingly different colours to the usual blue–black ones. With my family in Scotland I found myself discussing the colours of different lobsters that fishermen caught. One fisherman mentioned how the colour varied according to the depth of the sea water where the lobster lived. I looked up the penetration of light through sea water versus fresh water and at different depths. I also found a publication which mentioned that octopuses, another predator of lobsters besides humans, are colour blind [4]. This all seemed to me a beautiful area of biology, as was interacting with the topics of the physics of light and of coloration chemistry of carotenoids. Around this time, I learnt that there was a Society of Biology (now the Royal Society of Biology). I applied to be a member and was accepted. Some years later they wrote to me and asked if I would like to apply to be a Fellow. They provided me with names of Fellows of their Society from the United Kingdom. I knew some of them and asked two if they would be my supporters and they agreed. So, I became a Fellow of the Society of Biology.

One day I noticed a book by Ernst Mayr entitled *What Makes Biology Unique?* [5]. It was a fascinating read. He did not really accept the importance of the discovery of the double helix of DNA and its ramifications for the molecular basis of inheritance. Rather, he saw all those types of studies as extreme reductionism. This is a similar view to the one I quoted from Ottoline Leyser at the beginning of this chapter: "The defining feature of biology during the past few decades has been figuring out details of the parts. But biological systems don't think they have parts" [1].

So, the feelings that I had, and still have, of having such an aversion to dissecting even an earthworm to see inside it seems to me empathetic with Ernst Mayr and Ottoline Leyser, whereby I much prefer keeping the biology whole, e.g. when growing my broad beans. Overall I find then that biology is such an incredibly broad subject ranging from molecular biology such as DNA, through to whole organism studies, animal behaviour and onto whole populations, speciation and so on. Biology is very, very, diverse.

REFERENCES

1. Turney, Jon (2019) "The puzzle of life" THES 21st February 2019, pages 44–45.
2. Schrodinger, Erwin (1944) "What is Life? The physical aspect of the living cell" Based on Lectures Presented in Trinity College, Dublin 1943.
3. Cianci, M., Rizkallah, P. J. , Olczak, A., Raftery, J., Chayen, N. E., Zagalsky P. F. and Helliwell, J. R. (2002) "The molecular basis of the coloration mechanism in lobster shell: β-crustacyanin at 3.2 Å resolution" *PNAS USA* 99: 9795–9800.
4. Messenger, J. B. (1977) "Evidence that octopus is colour blind" *Journal of Experimental Biology* 70: 49–55.
5. Mayr, Ernst (2007) *What Makes Biology Unique?* Cambridge: Cambridge University Press.

Part II

Scientific Career Choices: What to Do When Faced With . . .

5

Junctions

What does it mean to be clever enough? My father was a policeman and my mother a midwife. They had not been to university but clearly were intelligent people and my mother in particular was well-read. I had one cousin who had studied at Glasgow University (English) but who now lived in Italy, as his wife was Italian.

The 11+ exam is a major career junction placed before me when I was ten years old – the general idea being that more academic pupils would go to grammar school. Those pupils who failed the 11+ exam would instead go to a secondary modern school, as they were called, destined for a trade or apprenticeship, or factory or mining jobs. The 11+ exam was to determine if I should pass to my local grammar school, Todmorden Grammar School, Yorkshire, or not. I passed. Much later the grammar schools were dissolved by the state in favour of 'comprehensive schools' so as to avoid pupils at such a young age being funnelled one way into their local grammar school or into their secondary modern school.

Under different elected governments in the United Kingdom, this selection of pupils at age eleven for grammar schools became a political football. Socialist 'Labour' governments would systematically merge grammar and secondary modern schools. When Conservative governments were elected they would discontinue the merger policy. The whole education system became fractured, inconsistent from area to area, sometimes referred to as a 'postcode lottery'. I would like to think that under any school system I would have been able to fulfil my education and training. Much later I learnt that the spending on a pupil in a state school was a factor of 2.5 times less than a pupil in a private school. Such a dichotomy would be best rectified by bringing the state school funding per pupil up to private school levels. But that is then a matter of priorities between education and other calls on the government purse, such as defence, welfare and so on. My progress and development may, I suppose, have been 'faster' if I had been born to wealthy parents, who could have sent me to a private school. So, inequality is built into one's life. In the end, I have never felt impeded by my 'place in life', but enabled by it and very grateful, too, to my parents and the society into which I was born.

6

Crossroads

At any stage of one's career a crossroads, of larger or smaller importance, can occur. A big one for me was deciding which route to go for my PhD. This decision occurred in the final year of my undergraduate physics degree. I was undertaking an experimental final year project involving electron microscopy studies of thin metal films and imaging their dislocations. I got on well with my project academic supervisor, and his electron microscope technician, so that when I was offered a PhD to stay in that research group in the interesting area of surface physics I accepted. Meanwhile the final year physics option courses on astrophysics, biophysics and geophysics commenced in the spring term. The biophysics course lecturer was in Brazil on a research education exchange visit and was delayed in starting the biophysics course. I opted for astrophysics, looking forward to learning about cosmology, especially the origin of the universe. Six lectures in, we had covered in considerable detail the laws of planetary motion and so on. In this I did not find the excitement I was looking for and so when the biophysics course finally started I decided to attend a few of those as well. Both the astrophysics and biophysics course lecturers were outstanding teachers. But the biophysics quickly and fully engaged me. I decided to switch over to biophysics from astrophysics. It became steadily clearer to me that my interests and abilities in chemistry, which I had left behind in my school advanced level, were still very strong. I enjoyed biophysics enormously. One area that I found amazing was the protein crystallography and the work of Dr Max Perutz. My course lecturer handed out an article written by Dr Perutz for the magazine *New Scientist* entitled 'Haemoglobin the molecular lung' [1]. The article explained, with X-ray crystal structure analysis, the atomic level details of what happens when oxygen binds at each of the four successive haems in this protein, which is found in the blood. There is a cooperative action between the four haems mediated by the protein in which they are embedded. When one haem group has taken up oxygen, it becomes progressively easier for the second, third and fourth haems to do so. The cooperative action is superior to ordinary dissolution, where it is easy to dissolve a small quantity of oxygen in pure solvent but gets steadily more difficult as the dissolved oxygen concentration rises. This molecular level process takes place after inhaling oxygen in the lungs; each haemoglobin of 64,000 molecular weight carries its four oxygen molecules to the muscles of the body where the cooperative effect facilitates the release of all four oxygens. The use of this method of physics to understand a fundamental process in biology in terms of chemistry was, I thought, fantastic. It was mind-blowing, in fact. Max Perutz also had a marvellous way of explaining things. He described haemoglobin as both a *chemical machine* and as a *molecular amplifier.* For good measure he said: "How can the weak chemical reaction of the four tiny molecules of oxygen with the four iron atoms produce such a drastic rearrangement in this giant molecule, like four fleas that can make an elephant jump?"

I started looking into biophysics PhD opportunities in the United Kingdom, and specifically in protein crystallography. I went to the York University careers centre and found a thick book entitled something like *Careers Opportunities for Graduates.* I found an entry for the Laboratory of Molecular Biophysics at the University of Oxford based in the Zoology Department there. The Head of the Laboratory was Professor David C Phillips. I wrote to him enquiring about the possibility of a DPhil place. (In Oxford the PhD is referred to as a DPhil). I was invited down to an interview and met prospective supervisors. I found the interview with Professor Phillips quite daunting but he was impressed by my in-depth answers to his questions for which I had been excellently prepared by my undergraduate biophysics option course! He wrote to me offering me a DPhil studentship funded by the Science Research Council (SRC). I replied accepting the place indicating my preferred choice of project, and with that chose the supervisor (Dr Margaret Adams) offering that project.

Now, I cannot recall the precise timeline but I had in effect got myself into what might be termed an almighty mess. I certainly had clarity in my mind of what I wanted to do next – the project with Dr Adams. But it wasn't a simple *let's drive straight across at this crossroads, continuing with the plan of doing a PhD in surface physics!* Instead I was going to turn right for molecular biophysics, also at the same time confirming that I was not going to turn left in favour of astrophysics and cosmology. I then had the awkward meeting with my undergraduate project supervisor withdrawing my acceptance of the surface physics PhD place he had offered me. Naturally, he was not pleased at this late development, although I felt we were not long into the summer term.

Some 45 years later, I am still active in the field of protein crystallography but have diversified [2].

REFERENCES

1. Perutz, Max "Haemoglobin the molecular lung" New Scientist 17th June 1971, pages 676–678.
2. Helliwell, John R. "Integrating X-rays, neutrons and electrons in structural chemistry and biology" https://zenodo.org/record/1257089#.XESP-Wm6LtQ

7

Roundabouts

A roundabout has several exits. There may be some delay until the driver can turn off, depending on the map-reading passenger making up their mind. Of course, these days satellite navigation ('sat-nav') gives instructions about one's final destination at the start of a whole journey and, before a roundabout, tells you which exit to take. It would be nice if it were as simple as this in the real scientific life!

In my field of crystallography, the study of crystallisation of proteins is a major discipline in its own right; my two colleagues and I have written a scientific research monograph on the topic [1]. Some proteins are more straightforward to crystallise than others and some defy all efforts at crystallisation. There are numerous parameters that one can vary to induce nucleation of a few protein molecules, in order to initiate the formation and subsequent growth of a crystal. We can set up many different chemical conditions. Each parameter can be varied by a small amount. A very large number of combinations of conditions can be set up to find the best ones for growing nice crystals. Once we have crystals we use X-rays to measure their diffraction patterns which we can analyse to determine the three-dimensional atomic and molecular structure. What if we fail in these procedures? Do we go around the whole exercise again? To repeat exactly the same chemical conditions again would have no point as science, including crystallisation, is, or should be, reproducible. So there would be no point in repeating the whole exercise. It would be just 'going around in circles'! Firstly, however, there is the chance for an unexpected variation in the chemical conditions. If it is a person setting up the crystallisation trays rather than a robot there may be small changes, even 'errors'. If the person has a beard there may be a small hair that falls into one of the tray liquid drops and acts as a surface for protein crystal nucleation to occur. So, such a lack of reproducibility is in that sense an advantage in this case. In general, a lack of reproducibility is a bad thing for science.

Are there other options rather than repeat the whole circle of experiments again? One can try a method altogether different than X-ray crystal structure analysis. A method that has improved a lot in the last ten years or so in my field of structural chemistry and biology is that of cryo electron microscopy. In this case, a crystal is not needed and instead multiple images of a single molecular complex are studied directly. This new method has allowed the breaking out of going round in circles for those impossible-to-crystallise cases. Why were they impossible to crystallise? This could be for a variety of reasons. Firstly, although a homogenously pure sample of these molecules could be prepared, they can each be too flexible to crystallise in a simple, repeating, 3D array. Secondly, it may not have been possible to prepare a pure enough sample for crystallisation. So important is the improved method of cryo electron microscopy for the field of structural biochemistry and structural biology that its lead developers shared the Nobel Prize for Chemistry in 2017: Jacques Dubochet, Joachim Frank and Richard Henderson [2].

So, roundabouts or 'going round in circles' in a scientific life present one of the most challenging of all situations that a scientist encounters. Sometimes one has to park a topic, i.e. postpone further work on it until it becomes timely to reopen the work, for example when a new and improved method might come along.

A very different type of example is the controversial issues that can arise in science where politics steps in. In my area of science, synchrotron X-ray sources are a major player. Indeed, I spent a great deal of my career working on developing and applying these for X-ray crystal structure analysis. This was firstly at the United Kingdom's synchrotron radiation sources (with the first one, called 'NINA', as a user in 1976, then working at its successor machine 'SRS' from 1979 to 2008 and from 2008 with 'Diamond Light Source') and as a U.K. participant in the European project 'ESRF' from 1983 onwards. The SRS, which came on line in 1981, was closed down in 2008, having completed its lifetime to make way for the technically superior national X-ray source 'Diamond'. The proposal for this new national source was made in the mid-1990s from the laboratory that built and operated NINA and then SRS, the Daresbury Laboratory. The funding for Diamond was not forthcoming from the U.K. government. Then, in the late 1990s, The Wellcome Trust stepped in with a substantial capital investment to kickstart the new project. They argued that since the majority of their funded researchers were based in London and the South East the new synchrotron source should be located in that region of the country. Initially the government argued in favour of the existing site, Daresbury Laboratory, but not sufficiently to provide all the funds for it. Was it ethical to deny the proposers of the project from Daresbury Laboratory the chance to build and run the new X-ray source? The statement today from the Daresbury Laboratory's website [3] is:

> The ceremonial switch-off of the SRS was undertaken by Prof Ian Munro on 4th August 2008; it was open to all staff and was recorded for posterity. Those who had worked on the machine, in whatever capacity, were then invited to a party hosted by Ian Munro. Staff were also presented with a limited edition medallion

> to commemorate the contribution made to scientific discovery by the SRS at Daresbury. It was, however, deemed inappropriate to have a grand celebration of the end of the SRS as this was a hugely unsettled and uncertain time for many staff with transfers and redundancies. Indeed, staff who worked on the SRS can now be found at many other facilities round the world.

The research charity funding provided a breakout from the impasse, the going round in circles that had delayed for about five years the new national synchrotron Diamond being funded.

A final example of a going around in circles for scientists is Brexit. The European Union, with its various funding schemes, has played a massive role in British scientific research. These resources complement the national funding agency ones. Naturally, then, scientists are by a very large majority opposed to Brexit. Nevertheless we were but one constituency in the 2016 U.K. referendum on the matter which overall voted to leave the EU. The deal secured by the U.K. government in negotiation with the EU has been, at the time of writing (early 2019), repeatedly rejected in the U.K. Parliament. Will a solution to this particular going around in circles be found?

REFERENCES

1. Chayen, Naomi E., Helliwell, John R. and Snell, Edward H. (2010) *Macromolecular Crystallization and Crystal Perfection.* Published by Oxford University Press, with the International Union of Crystallography, Monographs on Crystallography Book Series.
2. https://www.nobelprize.org/prizes/chemistry/2017/summary/ Accessed January 22nd 2019.
3. https://www.astec.stfc.ac.uk/Pages/Synchrotron-Radiation-Source.aspx Accessed January 22nd 2019.

8

Traffic Lights

The red traffic light means of course that you and your vehicle are stopped. Amber warns that the light is about to turn red and you must brake, or it alerts you that it is about to turn green. When green you are allowed to move forward. In some countries there is no amber alert when the lights turn from red to green, in the United States for example. From this everyday example we have two situations in the workplace where the Red, Amber and Green (RAG) colour code can be applied.

Firstly, in project management, regular progress report meetings are held to monitor a project. Green means all aspects are proceeding as expected, namely to budget and to the expected schedule. Amber relates to some concern that must be monitored more frequently or, worse, indicates that more resources are needed. Red means the project must be stopped because, for example, it is flawed or its scope has widened considerably, so much so that it cannot proceed as envisaged.

Secondly, a person may feel that their career progression has hit a ceiling. It is like sitting at a red traffic light. What should one do in such a circumstance? There are likely to have been signs that one's career had been slowing down, equivalent to an amber traffic light. Green of course means one's career is proceeding expeditiously. What is the solution to the career ceiling challenge? To register for one's employer mentoring scheme is an excellent way to address the challenge. I discussed *how to be a good mentor* in my *Skills for a Scientific Life* book, Chapter 24 [1]. As part of the Manchester University Gold Staff Mentoring Scheme, I served as mentor to several mentees who were at a career ceiling. Mentor–mentee discussions are strictly confidential but in general terms I can describe some ways to approach the challenge so as to guide the mentee in their future efforts. A key step is a time allocation self-analysis. In a typical month, to what aspects of the job does the mentee commit their time? In universities, for an academic, there will need to be two analyses: one for during the teaching term or semester and the second one outside of teaching periods. The point of this is that the teaching term polarises completely the nature of one's working day. This time allocation analysis usually exposes several things. Most prominently, rather than time wasted as such, which is unlikely for an already accomplished person in their mid-career stage, will be disproportionate amounts of time given over to particular tasks. By disproportionate I mean undue amounts of time for the importance of the task. My *Skills* book addresses how to deal with this in Chapter 9 [2].

Let us come back now to the managing of a project and the use of the RAG colour code method. There can be a range of science project types where RAG monitoring can usefully be applied. The two most obvious are firstly the laboratory leader running a funded research grant project and secondly the development or purchase of complex science apparatuses or instrumentation. For unfunded research projects, their execution is likely to be so haphazard anyway that RAG monitoring would be difficult to apply, but not impossible. For educators, RAG monitoring in effect is applied, but not called as such, with undergraduate examinations and PhD progression monitoring involving the end-of-year transfer report.

In my experience as a scientific civil servant, a very challenging and necessary development that I wrestled with was the need for improved X-ray diffraction pattern detectors. I wrote out a brainstorming style of analysis which I published in the journal *Nuclear Instruments and Methods* [3]. The implementation of these ideas involved two main developments. The first was a collaboration with Dr J E Bateman based at the Rutherford Appleton Laboratory near Oxford, a sister laboratory to Daresbury Laboratory where I was based, for a multi-wire proportional chamber (MWPC) device [4]. The second was the purchase of a television-detector-based device from a company in the Netherlands (Enraf Nonius) based on a prototype of a device developed by Dr U W Arndt at the Medical Research Council (MRC) in Cambridge. Both projects developed Amber stage concerns. The MWPC design, so as to achieve sufficient spatial resolution, required a set of high voltage anode wires at a 1 mm pitch whereas previously a 2 mm pitch was the norm for that technology. The 1 mm pitch at high voltage led to occasional sparking across a pair of wires. The TV design was tested in a prototype and also revealed instabilities, this time from thermal variations. Both devices were made to work, eventually. Meanwhile, a new type of technology appeared, an image plate. I worked with a Japanese company (Rigaku Corporation, Tokyo) to implement this into a device. This technology had a slow style of readout of the image, being mechanically based rather than having the electronic readout inherent in the MWPC and TV devices. This type of device was offered for sale by several manufacturers, each with a different mechanical scanning mechanism. They worked nicely: no need for Amber or Red labels in the stage management of them in one's workplace. The slow readout of these image plate scanners was overcome with yet another type of technology, a charge-coupled device (CCD). Different manufacturers of these CCD diffractometers entered the arena. So, in a period of ten years, a massive change of capability from the photographic sensitive X-ray films that we used at the time I wrote my brainstorming article had occurred [3]. A further important phase of technology development has been the pixel detector device. This arose out of developments made at a government-funded laboratory in Switzerland, the Paul Scherrer Institute (PSI) (https://www.psi.ch/en). The PSI device has been commercially spun out by the company *Dectris* [5]. They have also diversified their product into several versions including one that works in vacuum, important for measuring diffraction patterns with X-rays in the wavelength range of 3Å to 5Å [6]. These pixel devices, to my knowledge, are always Green equipment purchases, never Amber or Red. This is a delightful situation to have reached after 20 years of struggles with the various types of X-ray sensitive devices that I described above.

REFERENCES

1. Helliwell, J. R. (2016) Chapter 24 in *Skills for a Scientific Life*. Boca Raton, FL: CRC Press.
2. Helliwell, J. R. (2016) Chapter 9 in *Skills for a Scientific Life*. Boca Raton, FL: CRC Press.
3. Helliwell, J.R. (1982) "The use of electronic area detectors for synchrotron *X*-radiation protein crystallography with particular reference to the Daresbury *SRS*" *Nuclear Instruments and Methods in Physics Research* 201: 153–174.
4. Helliwell, J. R., Hughes, G., Przybylski, M. M., Ridley, P. A., Sumner, I., Bateman, J. E., Connolly, J. F. and Stephenson, R. (1982) "A 2-D MWPC area detector for use with synchrotron *X*-radiation at the Daresbury Laboratory for small angle diffraction and scattering" *Nuclear Instruments and Methods in Physics Research* 201: 175–180.
5. https://www.dectris.com/ Accessed January 21st 2019.
6. https://www.dectris.com/products/specific-solutions/diffraction.

9

Obstacles

What is the most fundamental obstacle to progressing a career in science? It is that faced by an early-career researcher in getting that secure rung on the career ladder. It is the permanent post that becomes the be-all and end-all need that enables a postdoctoral researcher to lead an independent scientific life, meaning they can not only have ideas but can also secure funds for the staff, equipment, consumables, etc. needed to take those ideas forward.

Once that obstacle is overcome, scientists usually report that insufficient funds for scientific research impede their progress in science and discovery. This arises due to the research grant proposal success percentages, with all funding agencies worldwide being too low even for highly rated, international quality, proposals.

Let's take these two obstacles in turn.

My own career progression at the all-important postdoctoral-to-permanent-post transition was fairly typical. I secured a five-year post, a key aspect in this being my ideas for developing protein crystallography at the new U.K. synchrotron radiation source (SRS) which were based on my experiments on the previous machine NINA [1]. This post gave sufficient 'permanency' for me to apply for research grants which last for three years typically. My post as a lecturer in biophysics was jointly at Keele University and the scientific civil service (50%/50%) at Daresbury Laboratory near Warrington, about 45 miles apart and joined by good motorway connections. All was well, I thought, for the transition to permanency. Mrs Thatcher in 1983 stepped in, however, with severe funding cuts to U.K. universities, stating that they were far-too-comfortable institutions living off the public purse. They needed shaking up was her view. For Keele University, whilst not suffering the worst budget cuts (which hit Salford University), the cuts were severe. Keele decided that it would make their half of my post permanent. The scientific civil service decided that it would renew their commitment of the three-year renewable 50% portion of my salary, i.e. for those next three years. This, I decided on reflection, was not good enough for that stage of my career. So I applied for a post at Stanford Synchrotron Radiation Laboratory, in the United States. But the Daresbury Laboratory advertised a Senior Scientific Officer post with a specification well-matched to the experience I had by then gained [2]. I was offered and accepted the post. It was permanent. As well as developing further instrumentation and methods for my chosen field of crystallography [3] I could collaborate with and/or support users. This was undoubtedly successful [e.g. 4–6]. Anyway, after three years, I was exhausted from working hugely long hours, supporting users, developing a new instrument and undertaking new research methods ideas that I had. By coincidence, I was invited to apply for a lectureship at the physics department in York University, also a permanent post. I opted to make it 80% with 20% at the U.K. synchrotron at Daresbury. Again, travelling between York and Daresbury included good motorway connections. After three years in that post I was invited to apply for a professorship in structural chemistry at the University of Manchester to which I was appointed in 1989. Again, I chose to make this 20% at Daresbury. I remained at Manchester until I formally retired in 2012. The U.K. synchrotron moved south and the Daresbury synchrotron ceased operation in 2008. My interests had diversified to include my own structural chemistry and biology research programmes as well as helping to develop the European Synchrotron Radiation Facility and also neutron protein crystallography (both in Grenoble) and applying electron microscopy in my research [7]. My story here illustrates several common aspects, I think, for all scientists in overcoming the obstacle to a permanent post. I would highlight the role of the five-year post in transitioning to permanency. The second is that I was able to move places. This description above did not include the fact that my wife, a DPhil inorganic chemist, and I moved together. She took up different postdoctoral posts in Keele, York and Manchester Universities. She did not become permanent herself until much later than I, also at Manchester University. Indeed, why was she unable to secure a permanent scientific post until much later than I? It links with the discussion that I began in Chapter 1 (Section 3), about my very able final year undergraduate student who decided not to pursue a career in science because so few women succeed.

A third aspect to consider is the role of chance. In my case it was perhaps an unusual case of chance of Mrs Thatcher's attacks on the funding and autonomy of U.K. universities. Today, I think, there is no doubt anger widely shown by European early-career scientists at the Brexit vote to leave the European Union as it creates uncertainty for their futures here in the United Kingdom. The 2016 U.K. Referendum result on EU Membership was obviously finely balanced: it seems like a matter of chance in the end that the result was as it was. So, luck for the early-career researcher does come into it.

What about insufficient funding for scientific research? Typically, 25% of research grant proposals are funded. In my experience there are very few poor proposals; this is whether I recall my own proposals or all those that I have seen in my committee work or that I have assessed as a referee. I imagine that others, like myself, do not completely abandon their unfunded research proposal ideas. We reduce scope and lengthen the timeline to try and take those ideas forward. But the huge amount of time researchers spend on preparing those unfunded research proposals is surely a huge waste! University research managers naturally apply considerable pressure on their academic research staff to keep up the rates of submitting grant proposals. Similarly, universities are on the lookout for successful researchers with above-average grant proposal success rates. A situation develops akin to the footballers' transfer market!

REFERENCES

1. Helliwell, J. R. (1979) "Optimisation of anomalous scattering and structural studies of proteins using synchrotron radiation" Proc. of Daresbury Study Weekend, 26-28 January 1979 DL/SCI/R13. pages 1–6.
2. Helliwell, J. R., Greenhough, T. J., Carr, P. D., Rule, S. A., Moore, P. R., Thompson, A. W. and Worgan, J. S. (1982) "Central data collection facility for protein crystallography, small angle diffraction and scattering at the Daresbury SRS" *Journal of Physics E* 15: 1363–1372.
3. Helliwell, J. R., Papiz, M. Z., Glover, I. D., Habash, J., Thompson, A. W., Moore, P. R., Harris, N., Croft, D. and Pantos, E. (1986) "The wiggler protein crystallography work-station at the Daresbury SRS; Progress and Results" *Nuclear Instruments and Methods* A246: 617–623.
4. Rossmann, M. G. and Erickson, J. W. (1983) "Oscillation photography of radiation-sensitive crystals using a synchrotron source" *Journal of Applied Crystallography* 16: 629–636.
5. Acharya, R., Fry, E., Stuart, D., Fox, G., Rowlands, D. and Brown, F. (1989) "The three-dimensional structure of foot-and-mouth disease virus at 2.9 A resolution" *Nature* 337: 709–716.
6. Liddington, R. C., Yan, Y., Moulai, J., Sahli, R., Benjamin, T. L. and Harrison, S. C. (1991) "Structure of simian virus 40 at 3.8-Å resolution" *Nature* 354: 278–284.
7. Helliwell, John R. (2018) "Integrating X-rays, neutrons and electrons in structural chemistry and biology" https://zenodo.org/record/1257089#.XESP-Wm6LtQ.

10

Mountains

If we set an impossible goal and we liken it to the biggest mountain on Earth, Everest, then maybe this chapter title should be 'Mountain', singular not plural. If it is a decent mountain, one to test our every effort to conquer it, then one would be enough. We can imagine also that a Nobel Prize is a fitting accolade to recognise the climbing of a science mountain. Well, there are some people who have won two Nobel Prizes, so setting an ambitious goal twice is possible!

Max Perutz, whom I have mentioned elsewhere in this book as an inspiration to me in my final year physics option course on biophysics, was a person who took about 30 years to solve the 3D structure of the oxygen carrying blood protein haemoglobin. This required taking numerous individual steps in developing his chosen method, X-ray crystallography, towards finally applying it to such a large molecule of 64,000 molecular weight. It also required him to steadily acquire a detailed knowledge of the chemistry and physics of the haem and its embedded iron in two different oxidation states. He was awarded the Nobel Prize in Chemistry in 1962 jointly with John Kendrew. It has sometimes been said that in today's funding climate he would not have enjoyed such long and continued support. He was based at the Medical Research Council (MRC) Laboratory of Molecular Biology in Cambridge, an institution that has subsequently produced a large number of Nobel Prize winners even up to the modern era. So, continuity of funding is still possible. Can it be achieved in universities? Or, put another way, do university scientists get awarded Nobel Prizes? Well, indeed they do, but one can argue that their discoveries are usually of a more sudden nature. The core point is to do with research grant funding. To rephrase the question about Max Perutz and his Nobelist successors at the MRC Laboratory of Molecular Biology – did they need to be regularly, continually, applying for research grant support, i.e. every three years or so?

It does seem that a researcher today must chase multiple sources of funds, so that setting a single long-term objective that is a single mountainous ascent is not a viable approach. It would be better in that sense to set goals that are at a maximum of five years ahead.

There is another approach so as to still harness the goal-setting as big as a mountain, though. This is to define a theme within which there should be various sub-themes that are interrelated. Taken together in the fullness of time, these can be an obviously mountainous achievement.

The funding agencies, whom I am in danger of labelling as short-term-focussed, do in any case have the defence that they also define their own themes, which are often called 'grand challenges'. These then are themes set from upon high by the funding agencies. They tend to be societally focussed research. The United Nations itself has entered the arena with the setting of its Sustainability Development Goals [1]. These are focussed on practical and hugely important objectives like achieving clean water for all. I find that these sorts of initiatives are not so easy to translate into themes matched to one's particular science training.

Looking back to the start of my science research career, and where science is now, it has changed way beyond my imagination. The collective efforts of many scientists together have transformed the landscape beyond my wildest imagination of 45 years ago. If I were to select one area of huge change I would say it was computing power now versus then. If I would be allowed a second huge change to mention I would say it is the internet. Both areas have not only transformed all areas of science but all areas of society and people's lives. There have also been huge achievements from biology with genomics, from chemistry with pharmaceuticals, from physics with instruments and devices, and from astrophysics with our transformed understanding of the universe. Each one was a mountain that scientists have climbed. The collective imaginations of the communities of scientists have delivered a collective of mountainous achievements.

REFERENCE

1. United Nations (2015) "Sustainable Development Goals are the blueprint to achieve a better and more sustainable future for all. They address the global challenges we face, including those related to poverty, inequality, climate, environmental degradation, prosperity, and peace and justice" https://www.un.org/sustainabledevelopment/sustainable-development-goals/ Accessed January 26th 2019.

Part III

Examples of What Science Delivers or Will Deliver in the Future

In making a selection of *what science can deliver* I have picked a mixture encompassing biology, chemistry and physics, including examples with no obvious application and others with application. In a room full of scientists asked to state their own such selection, I doubt there would be much, if any, consensus! The chapters following are then, to say the obvious, just my personal selection. Given that science is an activity of humankind of our recent past, spanning only several centuries, it seems to me that these selections will change in the future.

11

With Physics We Can See Atoms

I was at a conference of the American Crystallographic Association (ACA) in May 1999 in Buffalo, New York and on the bus to go on an excursion. The bus driver asked us, "What do you folks do?" A senior member of the ACA on the bus promptly replied, "Oh, we are people who can see atoms". I hope the bus driver believed that answer because indeed we do see atoms in our daily research!

I imagine that most people would know that a glass lens can serve as a magnifying glass. They may have seen an elderly relative using a large one for reading small print in a book or newspaper. A science-minded young person may have received a microscope for a birthday and looked at a human hair, which is 25 microns in diameter, or an insect wing through its eyepiece. These two objects are visible to the eye and one can see a magnified image of them in the microscope. How far can such magnification be pushed using sunlight? The answer is not so far as to see atoms because the limit is set by the wavelength of the colours in the sunlight, between 3,000 and 7,000 Å (i.e. 0.3 to 0.7 microns). To see atoms, which are much, much smaller than a human hair, requires much shorter light wavelengths. Such wavelengths, around 1Å, are available with X-rays. But there is no lens to focus an image at such a short wavelength. Physicists came up with computer software and methods to convert a diffraction pattern, an unimaged picture, into an image. The software and 'phasing' methods replicate the function of a glass lens for visible light. Those methods with X-rays have required a crystal to obtain a sufficiently strong diffraction pattern. The huge number of molecules arranged in an orderly way in a crystal compensate for the weak scattering of an individual molecule. There are now over a million X-ray derived crystal structures, precisely determined, from the quite simple up to the very complicated molecular complex with hundreds of thousands of atoms. A more direct method is to use electrons, which can be focussed by electric and magnetic fields like a glass lens can be for sunlight. The electrons, like X-rays, scatter from the object 'sample' but because they interact strongly with the sample even single molecules or single atoms can give a measurable scattering pattern and that image can be directly focussed by a magnetic field 'lens'. This method is the basis of the 'electron microscope'. The complications using electrons are different from X-rays. The interactions of the electron beam are so strong that they make the sample jitter when the beam hits it and that blurs the image. New detectors, just like the ones that replaced photographic film in a person's optical camera, now allow movies of the electron microscope images, again just like one can now take movies of holiday or family events. Successive images in the electron microscope can then be aligned in the computer using software, largely compensating for the blurriness. Also the new detectors reduce the electron dose required to obtain an image in the electron microscope, important for imaging biological macromolecules as samples, which are easy to damage. These developments are so important to the study of the fundamental processes of biochemistry that the 2017 Nobel Prize in Chemistry was awarded for this work [1].

These improvements with the electron microscope have been steadily arriving over the years. For applications to biological molecules, cryo temperature freezing of the molecules has yielded high resolution details in the 'electron potential maps' that can be comparable with results from biological crystallography. The major advantage of the cryo electron microscopy is that a crystal isn't needed. My research colleagues and I have worked on understanding the coloration phenomenon in the shell of marine crustacea, such as the blue–black coloured lobster shell. We had solved the X-ray crystal structure of a portion of it [2], known as beta-crustayanin, whose molecular weight is 40,000 Daltons and contains two carotenoid molecules known as astaxanthin. Then we sought the structure of the full complex, the so-called alpha-crustacyanin, comprising eight of these beta-crustayanins with a total of sixteen astaxanthins. Unfortunately, the crystals of alpha-crustacyanin were not well-ordered enough to yield a comparable quality X-ray crystal structure of the beta-crustacyanin. We decided to use cryoEM to determine the alpha-crustacyanin structure. This required research grant funds to secure the necessary running costs for the time we used the special electron microscope plus research consumables and the specialist research staff time, i.e. salary costs. In turn the research grant proposal required sufficient preliminary results to convince the grants committee to fund our cryoEM study. These preliminary results were to come from the more traditional method of staining the alpha-crustacyanin molecules with uranyl acetate and imaging those in a more conventional electron microscope at lower resolution than would be available with cryoEM. That work is described in reference 3]. With the lower-resolution image we could still put into position eight copies of our high resolution X-ray crystal structure of beta-crustacyanin (see Figure 11.1).

Our research grant proposal to use cryoEM to determine the structure of alpha-crustacyanin was submitted for funding, containing our research plans and the preliminary results shown in Figure 11.1, but despite being rated internationally excellent quality science it was unfortunately too far down the priority ranking order to be funded.

Another aspect of *seeing atoms* is the challenge, in particular cases, of being sure of the identity of a particular metal atom. Of note is where there

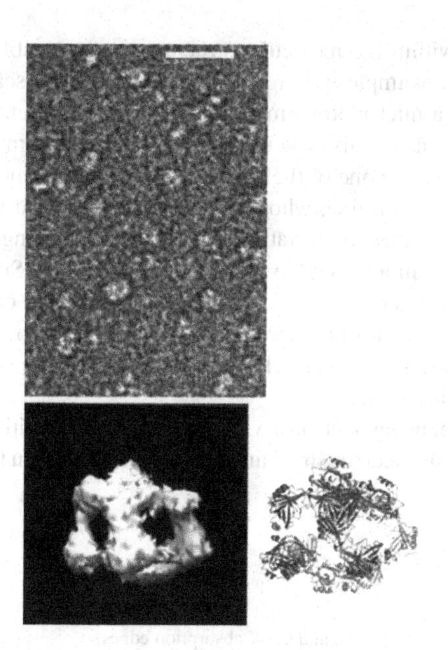

FIGURE 11.1 From reference [3]. Top: one representative negatively stained electron microscope (EM) image of α-crustacyanin from the 251 total images used to extract 10,021 alpha-crustayanin particle images used in the analysis. Scale bar = 50nm. Bottom left: one view of the EM three-dimensional reconstruction resulting from the analysis. Bottom right: the corresponding view of the final docked model comprising eight copies of beta-crustayanin from the X-ray crystal structure [2] placed into the EM image. The dimensions of alpha-crustacyanin were determined then to be 13nm by 14nm by 18nm. The EM analysis programme used was EMAN (Ludtke, S. J., Baldwin, P. R. & Chiu,W. [1999]. EMAN: semi-automated software for high-resolution single-particle reconstructions *J. Struct. Biol.* 128, 82–97). Reproduced with the permission of Professor Clair Baldock, who led the EM work, and the IUCr's *Journal of Synchrotron Radiation*.

are two metal atoms very close in atomic number in a molecular structure. A synchrotron radiation source which produces a continuum of X-ray wavelengths is especially useful for X-ray crystallography, in specific element location and various other ways such as having a very high X-ray intensity to compensate for the weak scattering of a sample the size of microns or for time-resolved studies [4]. A particular X-ray wavelength can be selected from the emitted spectrum with a 'monochromator' such as a robust and yet perfect silicon crystal. Then, by a rotation of that crystal to different angles to the X-ray beam, the monochromatised X-rays wavelength can be matched to a particular metal X-ray absorption edge involving the ejection of one of its core electrons. For two neighbouring metals adjacent to each other in the periodic table such as zinc and gallium the X-ray wavelength can readily be made selective to one or the other. With that method the identity of each metal within the molecular structure can be established unambiguously. Such an example was a study manipulating the scattering of zinc and gallium in a microporous material (Figure 11.2, reference [5]). There are many situations of this type in biology, chemistry or materials science. An example would be one of the first high-impact experiments I undertook on a synchrotron beamline, whose development I had led for such protein crystallography studies, SRS station 7.2, in that case distinguishing manganese and calcium in a vegetable protein (pea lectin [6]). Such tuned X-ray wavelength methods of diffraction analysis to be certain of the identity of metals at particular sites are incredibly powerful, more so than neutrons or electrons as scattering probes. This again emphasises the complementarity of the three probes [7,8].

An example of using neutron crystallography to see a critical proton in the chemical catalysis mechanism of an enzyme is described in the next chapter.

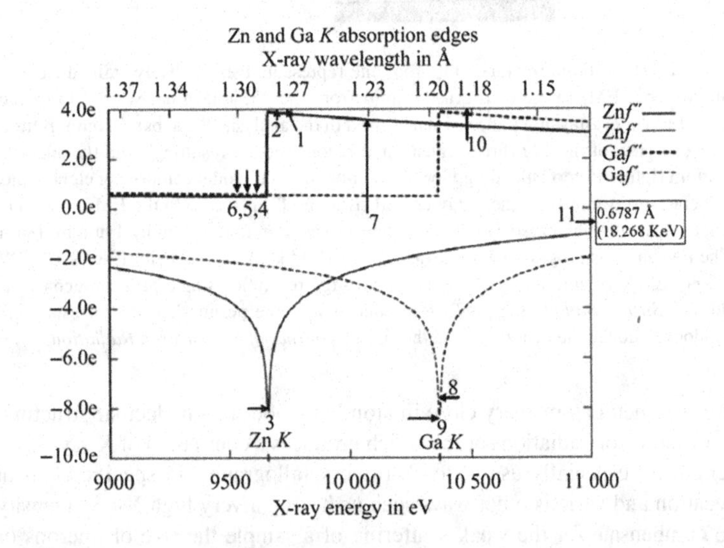

FIGURE 11.2 Absorption edges of zinc and gallium, with X-ray wavelength positions of the eleven single crystal X-ray diffraction datasets collected at the United Kingdom's Synchrotron Radiation Source Station 9.8 Daresbury shown by the arrow labels. These two curves show the real and imaginary parts of the X-ray scattering from an atom that vary with wavelength for each element. The real and imaginary components are necessary in a complex number description of amplitude and phase changes in the scattering, as the X-ray wavelength is varied. Reproduced with the permission of Dr Madeleine Helliwell, who led this study [5], and the IUCr's *Acta Crystallographica Section B: Structural Science, Crystal Engineering and Materials*.

REFERENCES

1. https://www.nobelprize.org/prizes/chemistry/2017/summary/.
2. Cianci, M., Rizkallah, P. J., Olczak, A., Raftery, J., Chayen, N. E., Zagalsky, P. F. and Helliwell, J. R. (2002) "The molecular basis of the coloration mechanism in lobster shell: Beta-crustacyanin at 3.2 Å resolution" *PNAS USA* 99: 9795–9800.
3. Rhys, N. H., Wang, M. -C., Jowitt, T. A., Helliwell, J. R., Grossmann, J. G. and Baldock, C. (2011) "Deriving the ultrastructure of alpha-crustacyanin using lower-resolution structural and biophysical methods" *Journal of Synchrotron Radiation* 18: 79–83.
4. Helliwell, J. R. (1992) *Macromolecular Crystallography with Synchrotron Radiation.* Cambridge: Cambridge University Press. Published in paperback 2005.
5. Helliwell, M., Helliwell, J. R., Kaucic, V., Zabukovec Logar, N., Teat, S. J., Warren, J. E. and Dodson, E. J. (2010) "Determination of zinc incorporation in the Zn substituted gallophosphate ZnULM-5 by multiple wavelength anomalous dispersion techniques" *Acta Crystallographica. Section B, Structural Science* 66: 345–357.
6. Einspahr, H., Suguna, K., Suddath, F. L., Ellis, G., Helliwell, J. R. and Papiz, M. Z. (1985) "The location of the Mn:Ca ion cofactors in pea lectin crystals by use of anomalous dispersion and tunable synchrotron X-radiation" *Acta. Crystallographica* B41: 336–341.
7. Helliwell, John R. "Integrating X-rays, neutrons and electrons in structural chemistry and biology" https://zenodo.org/record/1257089#.XESPWm6LtQ.
8. Helliwell, J. R. (2017) "New developments in crystallography: Exploring its methods, technology and scope in the molecular biosciences" *Bioscience Reports* 37: BSR20170204. Doi: 10.1042/BSR20170204.

12

Acceleration of Chemical Reactions by Catalysts: A Wonder of the Natural World

In Chapter 3 of this book I remarked how, when I was at school, I was struck in my chemistry lessons with wonder and awe at the periodic table of the elements. It rationalised this vast geography of the chemical elements and gave rational insights into what might chemically react with what. No doubt it is an excellent idea to have an International Year of the Periodic Table of the Chemical Elements as the United Nations science event of 2019 [1]. The event also marks the 150th anniversary of the discovery of the Periodic System by Dmitry Mendeleev (1834–1907) in 1869 in a presentation to the Russian Chemical Society. Mendeleev is pictured in Figure 12.1 when he visited Manchester on the occasion of the British Association Meeting in 1887.

When I say above that the Periodic Table *gave rational insights into what might chemically react with what* it masks the question of how quickly two elements might react. Also, might it be possible for a third party to such a reaction to speed up that reaction and yet not be used up in the reaction so that it can repeat its role over and over again? Such is the function of a 'third party' known as a *catalyst.* In everyday life when cooking a meal we know that heat increases the rate at which we can cook ingredients mixed together. Heat itself is not a *chemical catalyst* but it immediately gives us an idea that the time it takes to cook something can be speeded up. A catalyst then is a chemical entity which can speed up a reaction at any temperature including even room temperature. I am leading up to saying that life itself could not exist as we know it without the chemical catalysts in our bodies known as enzymes which allow life at body temperatures.

The understanding of how an enzyme works is transformed when its three-dimensional structure is determined by X-ray crystallography. The first such study was by a research team based at The Royal Institution in London led by Dr David Phillips in the early 1960s. Figure 12.2 shows a model of the structure of this enzyme, lysozyme, which breaks a bond in a bacterial cell wall. Why is it so called? Such a chemical breaking process is known as 'lysis' and so this enzyme became known as 'lysozyme'. The view

FIGURE 12.1 Mendeleev is pictured here when he visited Manchester on the occasion of the British Association Meeting held in 1887. The distinguished scientists featured in the photograph are as follows. Back row: Johannes Wislicenus (1835–1902), Thomas Carnelley (1854–1890), Henry Edward Schunck (1820–1903), Carl Schorlemmer (1834–1892) and James Prescott Joule (1818–1889); Front row: Julius Lothar Meyer (1830–1895), Dmitri Ivanovich Mendeleev (1834–1907) and Henry Enfield Roscoe (1833–1915). Figure courtesy of the University of Manchester. I thank James Peters, University of Manchester Archivist, and Emeritus Professor Jonathan Connor of the School of Chemistry, University of Manchester for making this figure caption as accurate as possible. Professor Connor also kindly provided this digital image.

direction in Figure 12.2 (top) shows the large cleft where the enzyme binds onto the bacterium. It is in the cleft that two reactive amino acids of the lysozyme sit waiting to do their chemical catalysis work. These are the amino acid residues aspartic acid 52 and glutamic acid 35 (Figures 12.2 middle and bottom, i.e. zooming in). They reside on either side of the chemical bond in the bacterial cell wall that is to be broken. Their protonation states (the presence or absence of a hydrogen) of their side chains determines their reactivities, one being protonated (glutamic acid 35) and the other not. After the chemical reaction, the protonation of glutamic acid is replenished by a proton in the water of which there are many available.

Lysozyme is present in our tears and mucus, as well as in a hen's egg white, and is a natural defence against bacterial attack, i.e. infections. Lysozyme is then an 'antibacterial' agent. Photographs of hen egg white lysozyme crystals are shown in Figure 12.3.

Figure 12.2 has a quite lengthy figure caption describing what has been found out about the nature of the three-dimensional structure of the lysozyme enzyme catalyst. Let's describe what needed to be done to find out those details of the protonation states of those two key amino acids. The protein had to be purified from the egg white. A crystal had to be grown of the pure protein (Figure 12.3). Actually the neutron study [2] used a different crystal form of lysozyme than that shown in Figure 12.3, used for the X-ray crystallography. For the protonation states, the crystal was placed in a probing beam of neutrons (see Figure 12.4). Also, because the probing beam is composed of neutrons, the hydrogens in the protein and the waters are best replaced by deuteriums, which have a strong and positive scattering for neutrons.

In Figure 12.4 the neutron beam of a quite broad range of wavelengths enters the apparatus from the left hand side. The single crystal is placed in the centre of the cylindrically arranged image plate at the right. The three arrows coming from the crystal are examples of diffraction rays arising from constructive interference of the neutrons passing through the crystal. We can think of a neutron as a particle, but here it is behaving like a wave. This is an interesting, if always puzzling, feature of quantum mechanics, which was mentioned in the 'What is physics?' chapter. The intensity of each and every diffraction ray measured with the image plate detector is estimated by computer software, in this case developed originally for synchrotron radiation X-ray Laue diffraction [4], as described in detail in [2]. This study by Bon et al [2] had several thousand measurements with the apparatus shown in Figure 12.4; specifically, 46,061 ray intensities were measured, which after averaging gave 8,976 independent observations. These measurements with a neutron beam deliver specific details of the hydrogen atoms bound to the protein and for the well-ordered water molecules bound to it. Prior to the neutron study, the X-ray crystal structure analysis, also at room temperature, described in [5], and its Protein Data Bank deposit 2LZT were used as the basis for the room temperature crystal structure analysis with neutrons of [2]. Neutrons are especially sensitive to deuterium atoms. This short description illustrates the complexity of the research undertaken to understand the role of a single proton in the catalysis of this enzyme, lysozyme. It has not yet allowed a time-dependent detailed analysis of the *catalytic process*. The crystallography data collection methods are speeding up to allow such studies, known as time-resolved diffraction [6]. This case study also illustrates the nature of interdisciplinary research, namely the bringing together of the scientific subjects physics, chemistry and biology. Furthermore, the computer software that I mentioned above is a complex suite of mathematical equations written in a computer programming language. The source of the neutron beams in this case come from a nuclear reactor. I also mentioned a synchrotron, a type of particle accelerator capable of producing X-ray beams. Both these types of beams, neutrons from a nuclear reactor and X-rays from a synchrotron, have a spectrum that is continuous over a wide range of wavelengths. This allows the software developed for processing

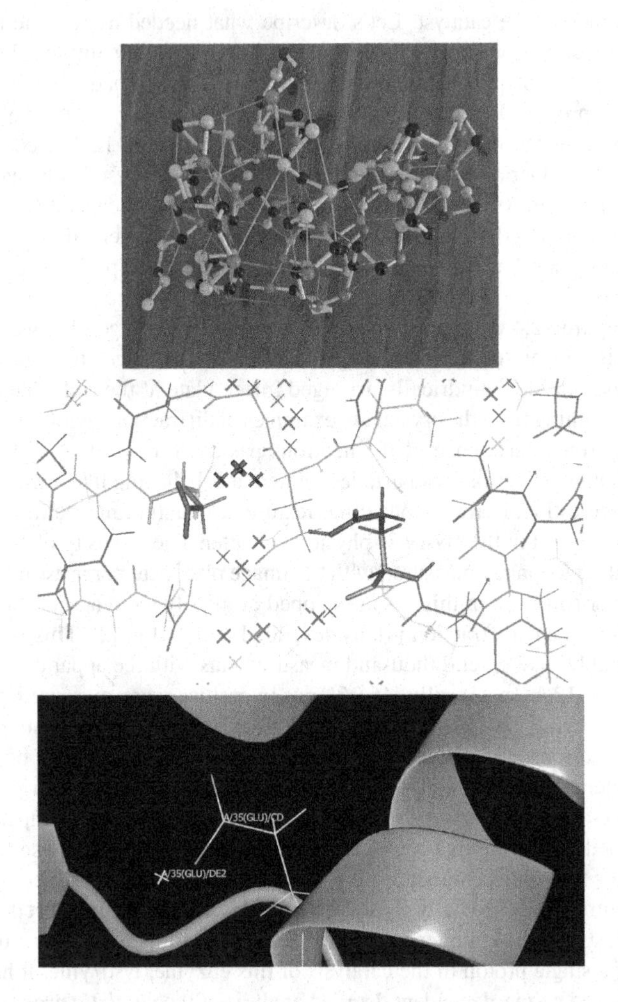

FIGURE 12.2 Top: the three-dimensional structure of the enzyme lysozyme shown as a simplified model with each one of its amino acids being a single bead with colours according to their chemical properties. The eight sulphur-containing residues, cysteines, have their sulphur atoms bonded in pairs, each making a disulphide bond, four in all, thereby significantly increasing the stability of the enzyme's structure. There is one other amino acid that contains a sulphur depicted here also, namely methionine, of which there are two in a lysozyme. This particular lysozyme is isolated from hen egg white. Other sources of lysozyme have different amino acid sequences, to a greater or lesser extent, known as their 'amino acid sequence similarity'. The dimensions of this protein are ~50x30x30 Å; 1Å is 10^{-8} cm. The Å is the preferred unit for interatomic distances of the International Union of Crystallography since, for example, a carbon hydrogen bond distance is 1Å and is therefore a very neat unit of interatomic distance. Middle: a zoomed-in view of the two catalytic amino acid residues with aspartic acid 52 (Asp52) on the left

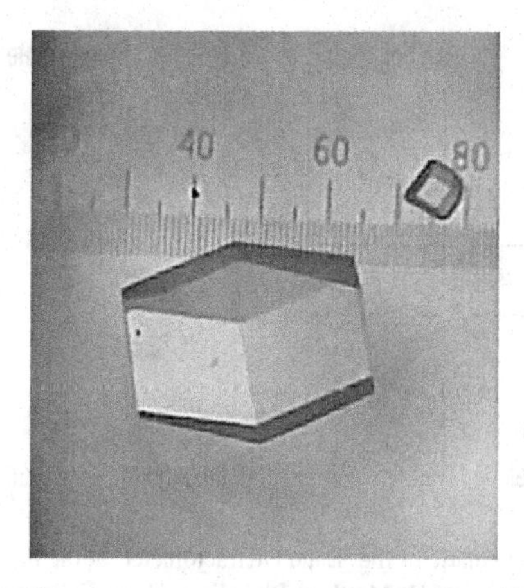

FIGURE 12.3 A single crystal of a hen egg white lysozyme can be formed by crystallising the purified protein in a high concentration of common salt, sodium chloride, which causes the lysozyme molecules to come out of solution to nucleate and grow quite large crystals. The larger of the two crystals is ~0.4mm in size; the magnification in the optical microscope is such that 80 divisions in the scale shown are 1mm. This picture includes a much smaller crystal at top right (~0.1mm in size) and also illustrates the variability of nucleation processes over time in the solution and at different places in the solution. This variability of crystallisation processes is much studied since it underpins the method of X-ray crystallography [3].

diffraction data for one to be adapted for the other. Both a nuclear reactor and a synchrotron require a great deal of sophisticated and precise engineering. I will explore in more detail the dependency of science on mathematics, the nature of interdisciplinary science and how science sits side by side with engineering in Part IV of this book.

FIGURE 12.2 *(Continued)*

and glutamic acid 35 (Glu35) on the right. Both have a chemically similar side chain, each with a carboxyl. The Glu35 is protonated, i.e. has a COOH group rather than the COO^- group that Asp52 has. The crosses are the bound water molecule positions including their oxygens and deuteriums. Bottom: a further zoomed-in view showing the Glu35 side chain with its proton (as a deuterium), labelled 35(Glu)/DE2 (Credits: The top image is a photograph of the lysozyme molecular model made by Beevers Molecular Models, Edinburgh. The middle image I prepared using COOT (P. Emsley, B. Lohkamp, W. G. Scott & K. Cowtan [2010]. Acta Cryst. D66, 486–501). The bottom image I prepared using CCP4mg (S. McNicholas, E. Potterton, K.S. Wilson & M. E. M. Noble [2011] Acta Cryst. D67, 386–394). The middle and bottom images are based on the neutron crystal structure analysis of C. Bon, M. S. Lehmann and C. Wilkinson (1999) Acta Cryst D **55**, 978–987 (reference [2]) and their data deposition coordinates file preserved at the Protein Data Bank, entry code 1LZN).

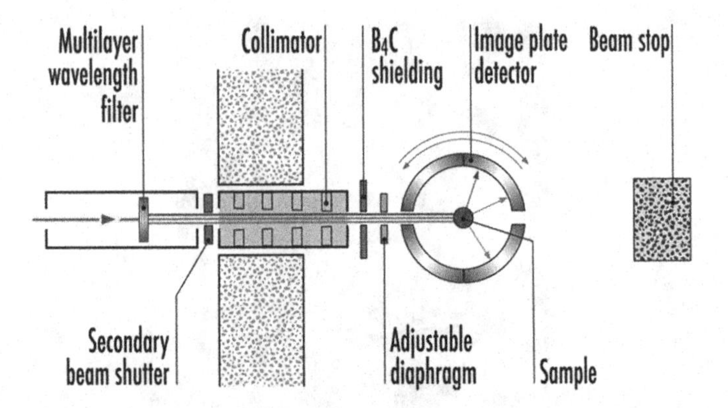

FIGURE 12.4 A schematic of the 'Laue Diffractometer' at the Institut Laue Langevin, Grenoble (figure courtesy of Dr Matthew Blakeley with credit for the image being to the Institut Laue-Langevin, to whom I am very grateful).

REFERENCES

1. https://iupac.org/united-nations-proclaims-international-year-periodic-table-chemical-elements/.
2. Bon, C., Lehmann, M. S. and Wilkinson, C. (1999) "Quasi-Laue neutron-diffraction study of the water arrangement in crystals of triclinic hen egg-white lysozyme" *Acta Crystallographica Section D: Biological Crystallography* 55: 978–987.
3. Chayen, N. E., Helliwell, J. R. and Snell, E. H. (2010) "Macromolecular crystallization and crystal perfection" (International Union of Crystallography Monographs on Crystallography) Oxford University Press International Union of Crystallography Monographs on Crystallography ISBN-9780199213252.
4. Helliwell, J. R., Habash, J., Cruickshank, D. W. J., Harding, M. M., Greenhough, T. J., Campbell, J. W., Clifton, I. J., Elder, M., Machin, P. A., Papiz, M. Z. and Zurek, S. (1989) "The recording and analysis of Laue diffraction photographs" *Journal of Applied Crystallography* 22: 483–497.
5. Ramanadham, M., Sieker, L. C. and Jensen, L. H. (1990) "Refinement of triclinic lysozyme: II. The method of stereochemically restrained least squares" *Acta Crystallographica Section B: Structural Science* 46: 63–69.
6. Helliwell, J. R. and Rentzepis, P. M. (Editors) (1997). *Time—Resolved Diffraction*. Oxford: Oxford University Press.

13

Understanding Colour: Paintings, Camouflage, Clothes and Cosmetics

In 2001, I visited the Prado Art Gallery in Madrid, having a spare morning break from my meeting on science facilities in Europe where I was representing the United Kingdom. A quite large number of paintings by Goya were on display. It was very striking how those from his later years were painted in brown colours versus the balance of colours in the earlier works. I thought this was very odd, of course, and imagined it was an eye defect that had developed with his advancing years. I learned that this was explained as due to his depressions from unexpectedly becoming deaf in middle age. This is an example of how our emotions are very much linked to our perceptions of colour in the world. A beautifully explained and illustrated book about the use of colours in painting by Marcia Hall has been recently published by Yale University Press [1]. It caught my eye that it was Yale University Press. I had visited Yale University in the mid-1990s to present a lecture at the retirement symposium of a colleague, Dr Arvid Herzenberg. In a spare afternoon I visited the Yale Center for British Art in New Haven, which I confess I was surprised to find, and a beautiful collection of art it indeed had, and has today (https://britishart.yale.edu/collections). My favourite artist is the French painter, George Seurat, who uses the pointillist technique. His painting *Sunday Afternoon on the Island of La Grande Jatte,* painted between 1884 and 1886, which I saw at the Art Institute of Chicago, is my favourite painting; a framed copy of it is in our dining room at home. That painting is a wonderful balance of colours and technique, and superbly conveys to me the mood of relaxation on a Sunday afternoon. Interestingly, the Art Institute describes the painting as having a busy energy…

In the laboratory, like Isaac Newton and his glass prism, we can split the white light of a lamp into the 'colours of the rainbow'. It is formally called a prism spectrometer. We can also use a spectrometer to measure the amount of light absorption in a liquid by measuring the amount of light transmitted into a detector and split that into an absorption spectrum. These instruments can separate the physical measurements from our perception and emotional moods. Our eye is obviously very sensitive to the colours of visible light. In my research on the molecular basis of the colour of the lobster's shell, I came across a publication that reported that another natural predator of lobsters besides humans, namely the octopus, is colour-blind [2]. So, what did the colour of a lobster matter to an octopus?

If we wish to evade a predator, whether we are soldier or animal, we need to blend in with the environment. Soldiers in a desert situation wear khaki clothes. Lions in the Kruger Park in South Africa have the brown of the surrounding grass land. But, then, why do zebras have black and white stripes? A favourite book, which my mother bought me when I was a child, was *Just So Stories* [3], which included tales such as 'How the leopard got his spots'. Camouflage was the answer as it enhanced his hunting success.

When we go to work we choose our clothes, and their colours, carefully. I would pick colours that would be different, usually not so bright, as for a dance night. In 1963, the movie *Cleopatra* starring Elizabeth Taylor came out. When I watched it I was struck by the beautiful colours of her eyebrows (deep black) and eyelids. Ancient Egypt was one of the first documented instances of the use of cosmetics. There was a cosmetic, an 'eye liner', called 'kohl' composed at that time of stibnite, antimony sulphide (Sb_2S_3). Due to the rarity of stibnite, the Egyptians also used galena, lead sulphide (PbS); it is actually one of the less toxic forms of lead.

REFERENCES

1. Hall, M. B. (2019) *The Power of Color: Five Centuries of European Painting*. New Haven: Yale University Press.
2. Messenger, J. B. (1977) "Evidence that octopus is colour blind" *Journal of Experimental Biology* 70: 49–55.
3. Kipling, R. (1902) *Just So Stories*. London: Macmillan.

14

The Universe Exists and the Big Bang 'Start' of the Universe: The 'Red Shift' and the Expansion of the Universe

I was at a conference in the Los Angeles Convention Centre in 2001. I had a spare morning or afternoon; I forget which. I decided I would visit the nearby Griffith Observatory Museum, which was a short taxi ride away. The exhibit at the time showed the hydrogen spectra of stars recorded by Edwin Hubble (1889–1953). These were compared with those on Earth, i.e. the laboratory spectra. According to the brightness of a particular type of star, which allowed its distance from Earth to be estimated, a method developed previously by Henrietta Leavitt (1868–1921), the spectra from different stars were placed one above the other according to their distance from Earth; those measured on Earth in the physics laboratory were also given as a reference. The spectral lines gradually showed a bigger and bigger shift in their position towards longer wavelengths. This is the so called 'red shift' effect. This means that the more distant the star is from Earth the greater its speed is in receding away from us. These are the observational, experimental, raw data that underpin our knowledge that not only is the universe expanding but also, if we were to count back in time, there must be a moment when all matter in the universe was concentrated at a single place. There must then have been an almighty explosion at the start which is referred to as the 'Big Bang'. Incidentally, we are not at the centre of that Big Bang. We too are receding from our neighbours as they are receding from us.

As I looked at these experimental spectra in the museum, I was overawed. It also felt somewhat chilling as I contemplated these bare details that mean so much to our current understanding of the universe. How it must have been for Edwin Hubble himself as he first placed his measured spectra one above the other for each star. I imagine he made an almighty shout of "Wow!", but I don't know of course.

When about to write this chapter, I went to the Griffith Observatory website but could not find any mention of that particular exhibition 18 years ago. I presume that somewhere in their store rooms and drawers these photographic plates are carefully kept.

Of course, even more fundamentally, as one contemplates the Big Bang moment, is the question: what brought this matter into being in the first place? Science is knowledge, as I mentioned at the outset of this book. But how can we know the answer to this question? A simple Google search of 'cosmology and the origin of matter' yields an immediately promising website [1] with the opening headline 'Cosmology: The Origin, Evolution & Ultimate Fate of the Universe, An Introductory Resource Guide for College Instructors'. Naturally enough, it doesn't in fact answer these questions and seems to be an open-ended website of many reading materials. Basically, we don't know enough about cosmology yet to know the origin of the universe, although we know much more about its 'evolution'. As for its 'ultimate fate', that part of the headline is also overambitious, I think, in terms of our knowledge at present. A book that I enjoyed enormously that describes all this, although a bit dated now, is by the physicist Steven Weinberg with the superb title *The First Three Minutes: A Modern View of the Origin of the Universe* [2].

REFERENCES

1. https://www.astrosociety.org/education/astronomy-resource-guides/cosmology-the-origin-evolution-ultimate-fate-of-the-universe/.
2. Weinberg, Steven. (1977) *The First Three Minutes: A Modern View of the Origin of the Universe.* 2nd edition 1993. New York: Basic Books.

15

Is There Life Elsewhere in the Universe? The Role of the Square Kilometre Array Radio Astronomy Project

The Square Kilometre Array radio astronomy project (SKA) states, on its website [1], that it is "an international effort to build the world's largest radio telescope, with eventually over a square kilometre (one million square metres) of collecting area". The radio telescope will have unprecedented sensitivity to weak signals and will thereby probe issues in astrophysics. The description continues:

> From challenging Einstein's seminal theory of relativity to the limits, looking at how the very first stars and galaxies formed just after the big bang, in a way never before observed in any detail, helping scientists understand the nature of a mysterious force known as dark energy, the discovery of which gained the Nobel Prize for physics, through to understanding the vast magnetic fields which permeate the cosmos, and, one of the greatest mysteries known to humankind... are we alone in the Universe, the SKA will truly be at the forefront of scientific research.
>
> **Quoted from [1]**

I listened to a progress report on the project at International Data Week held in Botswana in 2018 from the project's director, Rob Adam. His presentation provided a comparison of measured data rates with some other large scale projects. Firstly, as a reference, the total Google video streaming data rate is currently 2.2 gigabytes per second (gbps). Secondly, CERN in Geneva is currently producing data at a rate of 40 gbps. The SKA in its first phase will produce 300 gbps and in its second, full phase will produce 7,500 gbps. Such data rates, the largest ever seen, necessitate that instead of porting data to its principal investigators around the world the SKA has in its project scope for the data analysis to be handled centrally.

In its science objectives that I quoted above from the SKA website and heard directly in the lecture in Botswana, an obviously eye-and-ear-catching science objective is the one asking: "Are we alone in the universe?"!

So, in question time, I asked Rob Adam the following interrelated questions:

> If the SKA project finds evidence of life elsewhere in the universe, will it insist on a level of statistical certainty better than the typical 3 sigma, say 5 sigma? Secondly, how will this result be disseminated to the world? E.g. which journal and will there be a press release?

Rob Adam's answer surprised me: "Such a result would be passed first to governments". Why was I surprised? Well, as a publicly funded, i.e. taxpayer-funded, project it should, I think, open its results to the public and not solely to government representatives. To be fair to Rob Adam, though, I imagine he was balancing the public's right to know, as they pay for the project, and the possibility of mass panic in having access to such knowledge. An interesting aspect of the SKA planning on this again relates to their data archiving policy plans. At present they anticipate that the size of the primary experimental data files (the raw data) will be beyond what can possibly be archived. They will be able to archive their processed data. If life is found elsewhere in the universe with the SKA then those raw data that substantiate that will assume, I think, an immense importance.

REFERENCE

1. https://www.skatelescope.org/the-ska-project/.

16

Predicting Climate Change on Earth

That there might be a greenhouse effect on Earth was first postulated by Joseph Fourier in 1824 [1]. That it was due to carbon dioxide production was finally established in a series of studies by John Tyndall in 1859 [2]. Since that time, world population growth and the growth of the scale and number of industrialised economies has greatly increased the need for energy supplies. Fossil fuels have been a major source of these energy supplies. There is an international coordinated effort "to undertake ambitious efforts to combat climate change and adapt to its effects, with enhanced support to assist developing countries to do so. As such, it charts a new course in the global climate effort" [3]. Specifically:

> The Paris Agreement's central aim is to strengthen the global response to the threat of climate change by keeping a global temperature rise this century well below 2 degrees Celsius above pre-industrial levels and to pursue efforts to limit the temperature increase even further to 1.5 degrees Celsius. Additionally, the agreement aims to strengthen the ability of countries to deal with the impacts of climate change.

Without such concerted effort, dire consequences are predicted such as the melting of the polar ice caps and thereby a general rise of the sea level, leading to massive erosion of coastal land areas as well as major pressures on wildlife and on individual countries' economies.

The reports of the phenomenon and its acceleration through the last decades have led to vested interests of companies and countries ignoring warnings and sidestepping responsible limits. Worse still have been efforts to undermine the scientific reports with fake data or with publicity to label the science as fake news. In facing such hostility, scientists who study all aspects of the phenomenon and its effects may be tempted to go beyond providing objective facts and attempt advocacy to fight such climate change sceptics. I believe this is the wrong approach and instead I emphasise that providing objective facts should remain the paramount approach of scientists (see my critique of a book recommending advocacy in Appendix A2). In any case, one's personal decisions and behaviours, and the associated private morality embodied therein, as Dame Mary Warnock describes it [4], can emphasise the practical measures one can take to help reduce climate change such as:

- Ensure that as many journeys that one makes are by bicycle, not by car
- Fly less
- Take the train
- Use reusable bags when shopping instead of getting new plastic bags each time
- Refill water bottles from the tap instead of purchasing new bottles each time
- Eat less meat

Dr Gail Bradbrook, from the United Kingdom, is a leader in the founding of Extinction Rebellion, which organises mass protests to draw attention to these important climate change issues and to effect policy change [5]. Her important work on this has been profiled on BBC Radio 4 [6]. Greta Thunberg is a climate activist from Sweden who has brought major media attention to these issues via the practice of schoolchildren striking for the climate on Fridays.

REFERENCES

1. Fourier, J. (1824) "Remarques Générales sur les Températures du Globe Terrestre et des Espaces Planétaires" *Annales de Chimie et de Physique* 27: 136–67.
2. https://en.wikipedia.org/wiki/Greenhouse_effect.
3. https://unfccc.int/process-and-meetings/the-paris-agreement/the-paris-agreement.
4. Warnock, Mary (1998) *An Intelligent Person's Guide to Ethics*. London: Gerald Duckworth & Co Ltd.
5. https://extinctionrebellion.org.uk/.
6. https://www.bbc.co.uk/programmes/m0001lvz.

Part IV

Science and Mathematics: Across the Disciplines and Side by Side with Engineering

17

Science and Mathematics: Newtonian Dynamics and Molecular Dynamics

Mathematics is essential for science. This seems so obvious to me. This is especially true in physics, in the theoretical and physical branches of chemistry and also in the applications of statistics in hypotheses-testing in biology. Things that seem so obvious mean that they can be overlooked. In one department where I worked, the entry requirements by subject for incoming students did not include mathematics as essential. My campaign to require this was not successful, even though similar departments at other front-ranked universities like mine had this requirement. Let me emphasise the importance of mathematics in science with just a few examples.

I trace the thematic origin of one equation in physics, Newton's *force equals mass times acceleration* equation of dynamics, familiar to anyone driving a car or moving an object. As an application in chemistry and molecular biology it is known as molecular dynamics.

In 1666, Isaac Newton formulated the equation:

$$F = ma \tag{17.1}$$

i.e. force=mass multiplied by acceleration.

It wasn't a derivation, as is common in mathematics; his was a mathematical formulation, testable of course by experiment. We can readily understand what the equation (17.1) is summarising in its neat and succinct mathematics. A larger force is needed to shift a larger mass for the same acceleration, or a larger force for two identical masses is needed to accelerate one mass more than the other. Note that mass isn't the same as weight, as the television views of an astronaut in the weightless conditions of the International Space Station readily show us. Since the direction of the force and the direction of the acceleration also need to be described, equation (17.1) above must be recast in vector form:

$$\underline{F} = m\underline{a} \tag{17.2}$$

So, in equation (17.2) I have bolded and underlined both *F* and *a* to indicate that they are vectors and not scalars. Mass, however, is not a vector but a scalar quantity.

Newton's *force=mass multiplied by acceleration* equation features right at the beginning of my school advanced level set book [1]. I still treasure this book. It comprises 1,118 pages and I studied it avidly for two years at school before I entered university to study physics. Newton's equation was just one example of the feast of physics, carefully linked with the relevant mathematics, that is in this marvellous book.

Much later, when I was a professor of structural chemistry, one of our very top undergraduates, Gail Bradbook (whom I mentioned in Chapter 16 in her role as a leader in the climate change group Extinction Rebellion) approached me, saying she would like to do a PhD with me. It is a great highlight in the academic year when students express such a wish, especially an undergraduate student of obviously such a high ability. But, she said, she wanted to combine her planned PhD with the professor of theoretical chemistry, Prof Ian Hillier. This was precociousness indeed from her. With our respective experience, Prof Hillier and I could tailor a project for Gail Bradbrook's plan. We devised a project about the binding of different sugars to a protein that my laboratory was studying by X-ray crystallography. The particular protein was called concanavalin A, isolated from jack beans (as in the fairy tale *Jack in the Beanstalk*). Gail would firstly refine the crystal structure of glucoside bound to concanavalin A by using the measured X-ray diffraction data. This refinement procedure of X-ray crystal structure analysis in itself is underpinned by a mathematical equation known as the structure factor equation, where the calculated diffraction is given by:

$$\boldsymbol{F}_{hkl} = \Sigma f_j e^{2\pi i(hx_j + ky_j + lz_j)} \tag{17.3}$$

Where f_j is the atomic scattering factor for X-rays of the jth atom in the protein, x_j, y_j, z_j is the coordinate of that atom. $\boldsymbol{F}_{hkl}$ is the measured scattering for one of around 50,000 measured X-ray diffraction observations. Σ is a Greek symbol used to represent a summation of terms, in this case a summation of each atom's contribution of its X-ray scattering to the whole X-ray diffraction pattern of the crystal sample. The protein contained around 10,000 atoms, and bound a sugar molecule known as glucoside. The molecular model is best matched in the computations via equation (17.3) whereby the calculated diffraction is matched as closely as possible to the experimental, i.e. measured diffraction.

An earlier study by a previous PhD student of mine, Jim Naismith, now a professor and FRS, as well as Director of the Rosalind Franklin Institute at the Rutherford Appleton Laboratory near Oxford, refined the crystal structure of a mannoside bound concanavalin A at an identical X-ray diffraction

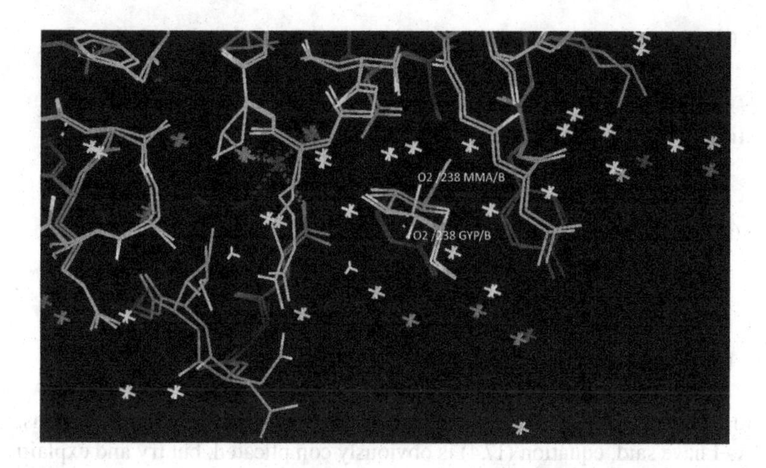

FIGURE 17.1 View of the sugar-binding site of the concanavalin A protein, zoomed in to see the two sugars studied in references [2] and [3] by X-ray crystallography superimposed. The chemical differences between the glucoside and the mannoside sugars is at one position labelled with the OH vertical in mannoside and horizontal in glucoside (the labels, respectively, are O2/238 MMA/B and O2/238 GYP/B). The crosses are bound water molecules. The two crystal structures are retained at the Protein Data Bank with reference codes 1GIC (the protein with glucoside) and 5CNA (the protein with mannoside). This figure was prepared using Coot (Emsley, P., Lohkamp, B., Scott, W. G. & Cowtan, K. [2010]. *Acta Cryst.* D66, 486–501).

resolution [2] as the glucoside. Gail could then compare her new crystal structure with this previous one (see Figure 17.1).

So, where did theoretical chemistry fit into this? The core question was: could we relate these two crystal structures to the thermodynamics of the binding of each individual sugar as measured by calorimetry, namely the heat changes on binding? Although tiny, and the method in fact is called microcalorimetry, such heat changes are measurable. Gail developed a 'master equation' (17.4):

$$\Delta\Delta H_{m/g} = \Delta\Delta H_{desolv.S} + \Delta\Delta H_{conf.S} + \Delta\Delta H_{conf.C} + \Delta\Delta H_{inter} + \Delta\Delta H_{rot/trans.S} + \Delta\Delta H_{solv.CS} \quad (17.4)$$

This equation (17.4) is obviously quite complicated. To explain it, we considered that the difference in binding enthalpy for mannoside (m) and glucoside (g), on binding to concanavalin A (C), as due to a combination of the following terms in the equation:

(i) A difference in the perturbation of water around the sugars (S) on complexation with the protein ($\Delta\Delta H_{desolv.S}$).

(ii) A difference in the changes in configurational enthalpy for the sugar (S) and/or the concanavalin A protein (C) ($\Delta\Delta H_{conf.S}$, $\Delta\Delta H_{conf.C}$).

(iii) Different interactions within the mannoside complex with the protein compared with that of glucoside with the protein ($\Delta\Delta H_{inter}$), which may be evident from the crystal structures.

(iv) A dynamical motion of the sugars within the active site leading to different average interactions for the two sugars ($\Delta\Delta H_{inter}$ here taken as the average over an ensemble, and $\Delta\Delta H_{rot/trans.S}$ the contribution of rotation and translation to the enthalpy difference).

(v) A difference in the solvation of the complexes ($\Delta\Delta H_{solv.CS}$).

In this study [3], we assessed the importance of each of these contributions. As I have said, equation (17.4) is obviously complicated, but try and explain that in just words alone! This research study was fundamental science but with significant applied science relevance. Saccharide binding to proteins is a key feature of many biological processes. Insights into such binding processes are crucial to the understanding of many disease processes. This study [3] was also one of the early ones to bring together multiple methods to understand the core process of protein ligand binding in biological systems. The mathematical equation (17.4) was at the heart of trying to guide our thinking as to what was going on. I can refer to the whole subject area, the combination of structure and theory of biological molecules, as molecular biophysical chemistry. Point (iv) above, which considered a possible dynamical motion of the sugars within the active site leading to different interactions for the two sugars, proved to be pivotal. We applied Isaac Newton's equation (17.2) of 1666 in what is referred to as 'the method of molecular mechanics'. The atoms in the protein with its bound sugar are individually subject to forces. So, equation (17.2) could be used to calculate the trajectories of the atoms in the protein and sugar. Then, based on those, snapshots with time of the molecular dynamics were obtained [3]. These were very, very large calculations, which required around a week of computing on one of the most advanced computers of the time. The total length of time simulated was around 0.5 nanoseconds. The simulation time step, each with a new calculated structure, was 0.5 picoseconds. The complete simulation comprised 500 sets of new *protein with sugar* coordinates. Analysis of the molecular dynamics sequences by Gail revealed transiently forming hydrogen bonds between atoms of one of the sugars versus the others. We had managed then to reveal a dynamical picture of the protein and its sugar interactions. This study [3], with its combination of experimental measurements of X-ray diffraction and microcalorimetry, along with the theory and mathematics needed to calculate the molecular dynamics, has obviously had a wide impact as judged by the 110 citations it has received. The connection to Isaac Newton and his equation (17.2) of 1666 to our molecular mechanics study that we published in 1998 shows how new science builds on the science that has gone before. And, oh yes, it also shows how mathematics is a core part of what we do as scientists.

REFERENCES

1. Nelkon, M. and Parker, P. (1974) *Advanced Level Physics.* 2nd edition. London: Heinemann Educational Books.
2. Naismith, J. H., Emmerich, C., Habash, J., Harrop, S. J., Helliwell, J. R., Hunter, W. N., Raftery, J., Kalb, A. J. (Gilboa) and Yariv, J. (1994) "Refined structure of concanavalin A complexed with methyl alpha–D–mannopyranoside at 2.0Å resolution and comparison with the saccharide free structure" *Acta Crystallographica Section D: Biological Crystallography* 50: 847–858.
3. Bradbrook, G. M., Gleichmann, T., Harrop, S. J., Habash, J., Raftery, J., Kalb, A. J. (Gilboa), Yariv, J., Hillier, I. H. and Helliwell, J. R. (1998) "X–ray and molecular dynamics studies of concanavalin A glucoside and mannoside complexes: Relating structure to thermodynamics of binding" *Faraday Transactions* 94(11): 1603–1611.

18

Science across the Disciplines: Curiosity Respects No Science Subject Boundaries

Challenges put to scientists from society, as well as those arising from our own scientific curiosity, do not always, in fact rarely, respect the traditional science subject boundaries. One can try to cover all the bases in following one's curiosity and develop skills across every science discipline. However, funding agencies posing grand challenges, and offering monies for research that one can apply for, tend to require proposals to have a team of experts each with a track record of successfully tackling interdisciplinary goals. It is worth mentioning that interdisciplinary challenges, and the research undertaken to overcome them, are a different category of research. Such research also doesn't suit everyone.

It is obvious from my biographical summary at the start of this book that I have endeavoured to develop my skills across each of the traditional science disciplines of physics, chemistry and biology. If *The Times* were to ask me what label I would use to describe myself in its featured 'Birthdays today' column, what would I reply? My professional activity is as a 'crystallographer' and this has taken me from my physics training across many scientific topics, through to a professor of chemistry, including in recent years even into studying compounds for medical imaging and for anticancer medicine. My first 'tenure track' academic post was as a lecturer in biophysics in a physics department, obviously a post where I was astride the two science subjects of physics and biology. I have endeavoured to serve both the crystallography and the biophysics science communities, each of which require an interdisciplinary approach. In crystallography I have served the International Union of Crystallography (IUCr) in several leading roles including representing the IUCr at the International Council for Science's Committee on Data (CODATA). For biophysics I led the U.K. Delegation to the Congress and General Assembly of the International Union of Pure and Applied Biophysics in New Delhi in 1999. These various roles emphasised the need for both a broad personal perspective of science subjects and also a need for good and constructive dialogue across the disciplines from all global players in science. I come back to the importance of the International

Council for Science in the final part of this book. A marvellous description of science across disciplines is in a lecture presented by Dame Nancy Rothwell FRS in 2019 [1].

REFERENCE

1. Rothwell, N. the Speaker's Science Lecture presented by her at the state apartments of the Speaker of the House of Commons https://www.bbc.co.uk/iplayer/episode/m0005c1r/speakers-house-professor-dame-nancy-rothwell?dm_i=4ZQ5,5R2B,2BIUBI,KRJX,1.

19

Science Side by Side with Engineering

Engineering can be so important in a scientific life. As a boy and then a trainee scientist at university, I used to think of engineering as building road or railway bridges, such as those by the pioneering English mechanical and civil engineer Isambard Kingdom Brunel (1806–1859). The change in my opinion, that engineering was also instrumentation for science experiments, happened in 1979, when I visited the Stanford Linear Accelerator Center (SLAC) and saw the 70-ton detector on a swinging arm rotating about the colliding beams interaction point. The detector had to be very precisely positioned, despite being so very heavy and bulky. At the time I was (partly) employed by the Science and Engineering Research Council, emphasising that engineering and science are juxtaposed!

A major part of my career was leading the construction, commissioning and use of synchrotron beamlines and instruments for crystallography, mainly biological crystallography but also chemical crystallography. On the team was a mechanical engineer with strong support from electrical and electronic engineering. I was the lead scientist. I described applications of these beamlines in Chapter 11. There were some specific moments where I had to press hard on the mechanical design to ensure the instrument delivered its science aims. Later I led the setting up of the European Crystallographic Association's Special Interest Group for Instruments and Experimental Techniques, the IET SIG. Since the invention of the first stored programme computer, 'the Baby', in Manchester in 1948, arising out of the Colossus 2 at Bletchley Park for code-breaking in World War II [1], the use of software has become essential for controlling science apparatus as well as calibration for proper data measurements. This book [1] also emphasises the transition for the design and build of computers from the cogwheels of mechanical engineering to the valves of electrical engineering and which of course have been transformed via electronic engineering with transistors and integrated circuits.

Overall, as I emphasised in Chapter 12 of *The Whys of a Scientific Life*, how technology pushes and pulls forward scientific ideas is an amazing feature of the progress of science.

REFERENCE

1. Copeland, J. (2006) Chapter 9 in Colossus and the Rise of the Modern Computer pages 101 to 115 in *Colossus The Secrets of Bletchley Park's Codebreaking Computers* Edited by Jack Copeland. Oxford: Oxford University Press.

Part V

Science Is a Process

20

Successes Involve Striving to Avoid Failures in Science

Overall, science as a process has in the modern era a variety of ways of seeking to avoid failures. The most difficult failures to contain, perhaps, are when medicines work one way in the laboratory and another with clinical trial test groups or the population at large, for example in terms of side-effects. In the 1960s, a major tragedy occurred when the anti-cancer drug thalidomide was prescribed over-the-counter for nausea. This cure for morning sickness in pregnant women led to the thalidomide tragedy with birth deformities. In turn this led to improvements in drug regulations and control [1]. So, today, in the United Kingdom for example, we have the National Institute for Clinical Excellence (NICE) which was established in 1999 and provides evidence-based health care around quality standards.

More generally, what quality and checking procedures are in place in science? I described in Chapter 8 that in project management there is the traffic light method to monitor a project's progress, put it back on track or stop it if needs be. In the ethics section of Chapter 1, I described the problem for science of unforeseen consequences of discoveries. In that same chapter, I described how science can attain objectivity through the archiving of raw data from experiments and linking those data to the narrative of a publication, which is subjective in nature as it comprises the authors' words. Readers of a publication, be they editor, referee or anybody else, can directly check the calculations if the raw data are available. In rather few occasions, a reader may see the need to submit for publication a critique article of a published study and the original authors can offer a response. A journal is bound to publish both, again subject to refereeing. So, these are examples of what science as a process does to avoid failures. The fulcrum of this process involves data. A movement has developed to ensure that the data is 'FAIR', which is to say findable, accessible, interoperable and reusable. The most technical term is 'interoperable'. I will explain it. Data are produced, i.e. measured by a specialist, so that other like specialists will know how to reuse it, provided there are sufficient details, known as the metadata, of the data. Other types of specialist might be interested too and they should also be able to reuse those data. The data must therefore be *interoperable across disciplines.* When might this be needed? Let's take an example: public anxiety about nanomaterials and nanotechnology. This was inflamed by the novel *Micro* of Michael Crichton [2] as well as widely reported comments such as those supposedly made by Prince Charles of the 'grey goo nightmare' of nanotechnology, a terminology which he has denied using [3]. The International Council for Science commissioned a working group led by its Committee on Data (CODATA) to develop a uniform standards system for nanomaterials. I was a member of the working group as the Representative of the International Union of Crystallography [4]. This standardised system for characterising nanoparticles is a vital step for ensuring a proper description of the risks of nanoparticles and nanotechnology, side by side with the commercial and societal opportunities.

REFERENCES

1. Heaton, C Alan (2013) *The Chemical Industry.* Dordrecht: Springer.
2. Crichton, Michael and Preston, Richard (2011) *Micro.* London: Harper Collins.
3. https://www.princeofwales.gov.uk/speech/article-hrh-prince-wales-nanotechnology-independent-sunday.
4. CODATA-VAMAS Working Group On the Description of Nanomaterials (2016, June 30) "Uniform description system for materials on the nanoscale, version 2.0. Zenodo". http://doi.org/10.5281/zenodo.56720.

Part VI

A Trend: The Coming Together of the Sciences and the Social Sciences

21

The International Council for Science: A Very Important Event

An important global redefinition of *What is science?* is the merger in 2018 of the International Council of Scientific Unions and the International Social Sciences Council to launch the International Council for Science. The merged mission reflects an age where funding agencies more and more wish to see scientific research undertaken that is valued by society. This is a very significant step which this chapter explores.

In Part I of this book, I offered my insights into what science is by looking at the questions *What is physics? What is chemistry?* and *What is biology?* An important recent event, in 2018, which challenges those conventional categorisations is the merger of the International Council for Scientific Unions, or ICSU, ('the sciences') and the International Social Sciences Council, or ISSC, ('the social sciences') to form the joint International Council for Science. This is a seismic shift in the organisation of the world's science. The motivation for bringing this about is described in reference [1] and, as a measure of its endorsement by both organisations, "the final vote count in favour of the merger for ICSU was 97.6% and 90% for the ISSC". At the heart of the concept of the merger is that, more and more, 'the sciences' are asked by funding agencies, in turn pressed by governments, to undertake societally relevant research. Since the impacts of such research will be social then in one way or another the social sciences should be joined with 'the sciences'. In terms of core details, the International Council for Science is a non-governmental organisation with a global membership of national scientific bodies (122 members, representing 142 countries) and International Scientific Unions (31 members). The philosopher Dame Mary Warnock [2] compares and contrasts 'the sciences', which deliver 'facts', and the values of a society, i.e. that value judgements are not facts. I spoke up personally about the merger at meetings of the International Union of Crystallography, such as its General Assembly held in Hyderabad, India in 2017, and in support of the idea of the joining up of 'the sciences' and 'the social sciences'. I believe that it offers a methodology for bridging 'facts' and 'values'.

In its plans, the International Council for Science envisages a focus on achieving the global sustainability development goals of the United Nations [3], which state:

> The Sustainable Development Goals are the blueprint to achieve a better and more sustainable future for all. They address the global challenges we face, including those related to poverty, inequality, climate, environmental degradation, prosperity, and peace and justice. The Goals interconnect and in order to leave no one behind, it ís important that we achieve each Goal and target by 2030.

So, the whats of a scientific life in the coming years will be a different flavour than before, although the whys of a scientific life will not be, since, as I described in my recent book [4], 'the why' is a more permanent entity than 'the what', being an expression of curiosity. It is curiosity which will always be a key driver for why scientists do what they do [4]. Curiosity of scientists must be respected in this new world of the merged International Science Council. This is vital, I think. It will be good for all scientists to consider, for example, the application of statistics to our data whether it be considering if busier hospitals have higher survival rates or the need for a 'five sigma' proof of the Higgs boson in particle physics, just two examples described in the illuminating and well-written book on data and statistical science by David Spiegelhalter [5], which very nicely shows us what this merged scientific world will be like.

REFERENCES

1. https://council.science/current/press/worlds-leading-bodies-of-social-and-natural-sciences-to-merge-in-2018-becoming-international-science-council.
2. Warnock, Mary (1998) *An Intelligent Person's Guide to Ethics*. London: Gerald Duckworth & Co Ltd.
3. https://www.un.org/sustainabledevelopment/sustainable-development-goals/.
4. Helliwell, John R. (2018) *The Whys of a Scientific Life*. Boca Raton, FL: CRC Press Taylor and Francis Group.
5. Spiegelhalter, David (2019) *The Art of Statistics: Learning from Data*. London: Pelican Books.

Appendices

My Reviews of Books Regarding the Whats of a Scientific Life

Appendix A1: The Social Function of Science, *by J. D. Bernal*

Reference: John R. Helliwell book review of *The Social Function of Science*, by J. D. Bernal (1939), published by George Routledge and Sons, London. *Crystallography Reviews* (2018). **24**:4, 280–285.

An internet version of *The Social Function of Science* by J. D. Bernal is available at https://archive.org/stream/in.ernet.dli.2015.49995/2015.49995.Social-Function-Of-Science#page/n507/mode/2up

This book remains a classic work in the landscape of science and society. At the time of writing this book, J. D. Bernal (1901–1971) was Professor of Physics at Birkbeck College, University of London having moved there from Cambridge where he was Lecturer in Structural Crystallography. Bernal became a world recognized expert in his scientific field of crystallography as well as a world recognized polymath, known as 'Sage'. The aim that Bernal had for writing his book he describes in his Preface as 'the book will have served its purpose if it (shows that) the proper relation of science and society depends on the welfare of both'. The book is divided into two parts each with eight chapters. Part I is entitled 'What Science Does' and Part II 'What Science Could Do'. There are 10 appendices. I found the book a remarkable insight into the state of the World after the First World War and the economic slump, and in 1939 the grim reality of a world being plunged into the Second World War. These events I would surmise from the points made and emphases given by Bernal stimulated the book. But also his admiration for the organization of the Soviet Union, repeatedly arguing its benefits, is also fascinating. His many insights and profound thoughts more generally were a pleasure for me to discover in reading the book and of which I highlight various ones below. Naturally, I do not always agree with him. The book is available via the internet at the weblink given above but I also purchased a printed copy via Amazon books as the three charts in the book were not scanned properly on the web version. The charts are indeed very interesting offering Bernal's grand schema with charts 1 and 2 and his preferred solution, chart 3:-

- Chart 1: ORGANIZATION OF SCIENCE
- Chart 2: TECHNICAL PRODUCTION

- Chart 3: ORGANIZATION OF SCIENCE IN THE SOVIET UNION IN 1937

Chapter 1 is entitled 'Introductory' and its subheadings capture the tone and mood of the author. These are:-

- THE CHALLENGE TO SCIENCE: The impact of events; should science be suppressed?; the revolt from reason
- THE INTERACTION OF SCIENCE AND SOCIETY: Science as pure thought; science as power; disillusion; escape; the social importance of science; the scientist as worker; science for profit; the institution of science; can science survive?

Chapter 2 is an amazing summary of science in history. Chapter 3 is on the organization and financing of Science in Britain. On page 65 we learn, for example, that the percentage of the estimated gross domestic product (GDP) spent on research in Britain was at that time 0.1% versus 0.8% in the Soviet Union. Bernal observes on page 101 that 'As science advances the delicacy of the phenomena it observes continually increases, and this puts a premium on the use of more and more elaborate apparatus'. This is of course in the pre synchrotron or neutron central facility age (1939). On page 105 Bernal summarizes the tensions between teaching, research and administration workloads for the university academic, which is true to this day of course. Overall Bernal paints a fairly dismal view of the role of the university. Bernal's sharp criticism is more stinging about research in government laboratories and even more so for industrial research where, paraphrasing page 107, 'even scientific and technical books cannot be borrowed from libraries lest it might be revealed what a company is working on'. On page 110 he gives a swingeing critique of nine companies whose products were based on science, and which accounted for 3/4 of the industrial research in Britain, and 'whose directors basically treat their products as purely commodities and only 13 out of 114 have science qualifications themselves'. On page 113 he gives sharp criticism of the learned societies who have become 'publishing houses'. There seems to be no appreciation from Bernal of accurate publications i.e. properly refereed and edited and whose sales at small surpluses are ploughed back into the science community with student bursaries. Furthermore, on page 114, he comments on the insularity of each science discipline which is exacerbated by learned societies and the departmental structure of universities. Thus for example, 'chemists for a quarter of a century have failed to recognize that advances in physics and crystallography require not merely the revision but the complete recasting of the fundamental structure of their science' and on page 115 'One effect of this has been to hold back science just at those very places where its advance is most needed, the regions between recognised sciences'. Bernal states with obvious feeling that there are advantages of 'the existing anarchic state of science gives many opportunities to evade particularly obnoxious control'. The antidote to 'obnoxious control' he sees is one that is democratic with scientists of every grade of seniority involved in assigning resources, an elected approval committee one presumes. On pages 127–128 Bernal's political loyalties are evident when he states 'At present, outside the Soviet Union, production is everywhere carried out for private profit, and the use that is made of science is primarily determined by its contributions to profit'. From our modern perspective, we would note the collapse of the Soviet Union, including the symbolic tearing down of the Berlin Wall, rather than democratic capitalistic states, events which would have been interesting to discuss with Bernal.

At page 153 in the chapter VI of applying science in industry, including patents, we come to the heart of Bernal's jaundiced view of commerce:-

> The object of commerce in respect to the consumer is not to provide him with the best goods at the lowest prices, but the cheapest goods at the highest prices that can be maintained by restricting competition ... The position can only improve if we can concurrently develop science and redirect productive processes for welfare and not for profit.

Bernal fails in my view to see the power that the consumer wields when there is a choice of product to pick one over the other. Bernal concludes this chapter on page 159 with a firm statement: '*arguments against plenty under socialism are refuted by the actual experience of the U.S.S.R.*'. There are fascinating facts revealed in Bernal's obviously very diverse reading. On page 161 we learn that '*only 12 out of the 75 most important inventions made between 1889 and 1929 were products of corporations' research*'.

In Chapter VIII Bernal kicks off on page 191 with a very moving comment:-

> Science has been from the start international in the sense that men of scientific temper even in most primitive times were willing to learn from others in different tribes or races ... The history of the main stream of modern science, from Babylonian to Greek, from Greek to Arab, and from Arab to Frank, shows how effective this process has been.

The chapter goes on, however, to mainly dissect through a mix of fact and Bernal's candid opinions the science efforts in a variety of different countries. Some of these one might say, perhaps from our perspective in the politically correct world of today, to be almost libellous, if libel can be applied to a whole nation. From page 215 Bernal charts in a strict evidential way the alarming brutalities against certain peoples notably 'The persecution of the Jews' including unbelievable writings of the anti-semitic Professor Stark in the journal *Nature* no less, which Bernal quotes verbatim and in even more extreme form in the anti-semitic writings of Stark in the organ i.e. newsletter of the German SS. The target of Stark's disgusting attacks is what Stark called the 'dogmatic approach to physics', as opposed to the pragmatic approach, as exemplified through the use of ideas such as from Einstein, Schrodinger, Born or Heisenberg. This section culminates in Bernal quoting from Hitler's *Mein Kampf* 'The State must throw the whole weight of its educational machinery not into pumping its children full of knowledge, but into producing absolutely healthy bodies … scientific training follows far behind'.

Bernal in the next section of this chapter goes on to extol the virtues of the masterplan of the Soviet state. He quotes extensively of this masterplan with its 10 focal points for scientific research. These sound quite similar to directed science research programs in the democracies today but the Soviet one of 1938 goes on to state 'It does mean that the lesser research problems (to the ten priorities enumerated) will be subordinated to questions which are of vital need to the country as a whole'. In effect this is a near total prescription to directed research topic areas rather than the imaginative blue skies creative thinking of scientists. The core defence, and enthusiasm, of Bernal of the Soviet master plan approach is 'Thus industry serves to present science with new and original problems. At the same time any fundamental discoveries made in the universities or the Academy are immediately transmitted to the industrial laboratories'. The central Soviet bureaucracy is presumably the controlling beneficence of this scheme. I betray here of course my continued enthusiasm for the science of the modern world democracies, after more than 40 years directly involved in it.

Part II of the book is entitled 'What Science Could Do'. Naturally, this immediately leads Bernal into describing training of scientists, starting with the topic of school science but also in the approaches of society to its citizens learning of science. He makes a sweeping set of suggestions for curricula revision in all the science subjects. There is a consistent lament: the need for new science knowledge to be incorporated promptly in courses and also a keen attention to bringing in aspects of other science disciplines, such as biology into chemistry or physics courses.

Chapter X addresses the 'reorganisation of research from first principles' as envisaged by Bernal. These principles he states as follows:

> We always have to keep in mind two main considerations. The first is that research is carried on ultimately by individuals and that therefore the conditions of the individual research worker require primary attention. The second is that as research is carried on for the benefit of humanity as a whole, it requires the most effective coordination of the work of the individuals … The main problem

> is the reconciliation of the needs of organisation for the whole with the freedom of the individual.

This is to my view an insuperable problem. For example, Einstein was surely motivated only by a basic question such as what if the speed of light is finite. Under Bernal's envisaged regime this is not countenanced. The same difficulty lies with the modern grant proposal impact statement, which cannot be envisaged except in the vaguest of terms or in an applied science grant proposal in the most precisely defined, basically obvious, terms. Bernal continues his overview and recommends a reorganization of journals in Chapter XI entitled 'Scientific Communications' not least to get away from the 33,000 science journals at his time of writing his book, and with new titles being constantly introduced (see page 292). He also envisages a public library of science and in effect open access to all readers, including the public. This is elaborated on in full in his book's Appendix VIII. Bernal was certainly a visionary ahead of his time!

In Chapter XII Bernal addresses 'The Finance of Science: Science and Economic Systems'.

Bernal states that 'the chief abuse in the present financing of science is not its rigid character but its extreme variability'. He prefaces this remark by saying basically that either money is not available when it is needed or when there is a surplus labs do not wish to return the money and so it is wasted. On page 313 he takes up the cause of the scientific worker. He declares that 'to be able to do his best for science … he needs security of employment, adequate leisure and appropriate status … The policy adopted in France of a definite graded profession with possibilities of exchange into teaching or administration offers an ideal solution'. On the question of pay Bernal puts in a plea for an increase in pay for junior scientists whilst defending the pay of a professor, such as himself.

Chapter XIII is, excitingly, entitled 'The Strategy of Scientific Advance: Can Science be Planned?' The core new idea offered by Bernal within this theme is within a delicious sentence 'Here conscious direction should prove far more flexible than the existing planless development of science'. By delicious I mean it seems to me rich in its self-contradiction! Chapter XIV is entitled 'Science in the Service of Man: Human Needs'. Bernal takes a long-term view on page 380 when he states 'The development of space navigation, however fanciful it may seem at present, is a necessary one for human survival, even though that necessity may lie a few million years ahead'. Chapter XV is entitled 'Science and Social Transformation: Social Conditions and Science'. This chapter develops a description of the two extremes, as we would say today, of Soviet communism versus German or Italian Fascism. Interestingly the middle way, which I would label as 'regular democracy', hardly comes into the logic of Bernal's description in this chapter, which is bizarre. But the huge forces that were Nazism or Soviet communism were one can readily assume the dominant ones of 1939, when this book was published. On page 398 is a section entitled 'Scientist as Ruler'. Bernal, whilst serious, does express the misgiving that 'most existing scientists are manifestly totally unfitted to exercising such control'. On pages 402–403 Bernal has a section entitled Science and Politics. He is clearly in favour of activism by scientists on issues whilst admitting that this has serious dangers. He recognizes that to go beyond neutrality a scientist risks being labelled tendentious. He allows one categoric exemption which is when the existence of science itself is threatened such as Galileo experienced from the religious establishment of his day.

Chapter XVI is entitled 'The Social Function of Science'. Again we have a sentence of piercing analysis from Bernal, on page 410, 'The starving of research of potential human value is but one step removed from the starving of man'. A disappointing statement on page 411 in the Science and Culture section is to do with the science method where he states that 'there is no science of science'. This is because he states that 'the making of discoveries lies outside the scientific method and is instead the operations of human genius'. Yes, I would observe that Einstein was a genius, of that surely there can be no doubt. He asked questions out of curiosity like 'what if the speed of light is finite'. This is but one of the types of the scientific method.

On page 415 we reach the flaw in Bernal's approach which he states and explains that 'Science is communism'. This denies the role of the individual scientist and of an individual's curiosity. Bernal senses this when he then states 'In science men have learned consciously to subordinate themselves to a common purpose without losing the individuality of their achievements'.

There are several Appendices where Bernal meticulously documents science research and its organization as well as financing, numbers of research workers and so on in the UK, France and the USSR. Appendix VIII, as mentioned above, is a visionary proposal for a radical alteration of publication by scientists involving a not-for-profit centralized publishing house instead of a scattered arrangement of publishers including for-profit ones. It is a vision of the Public Library of Science of today. It has a strong theme of harnessing the technology of microfilm rather than the more expensive typesetting. Appendix IX is a statement for an International Peace Campaign. This includes a proposed, inspiring, resolution 'We recognise that war is fatal to science, not only by breaking up its fundamental international character, but even more by destroying its ultimate purpose of benefiting the human race'. Finally, he makes a proposal for an Association of Science Workers, with basic needs articulated for their pay, work conditions and pensions. Also, he makes a key proposal for a wider influence of scientists such that 'The Association will work towards securing direct representation of scientific experts on all Government Commissions and Committees and upon all public bodies whose findings may affect the interests of scientists or the application of science to society'.

This book remains an amazing document.

Appendix A2: The Effective Scientist: A Handy Guide to a Successful Scientific Career, *by Corey J. A. Bradshaw*

Reference: John R. Helliwell book review of *The Effective Scientist: A Handy Guide to a Successful Scientific Career* by Corey J. A. Bradshaw (2018), published by Cambridge University Press. *J. Appl. Cryst.* (2018). **51**, 1259–1261.

Once one has embarked on a science career the question arises, how can one maximize one's effectiveness? This book by the very accomplished global ecologist Corey Bradshaw is devoted to explaining that. My own book *Skills for a Scientific Life* (Helliwell, 2016) to some degree creates a need to keep my observations in this review of the book by Corey Bradshaw as specific and thereby as objective as possible, but it also shows our joint interest in this topic. I had a similar situation with my recent review of the book *Scientific Leadership* (Niemantsverdriet & Felderhof, 2017; Helliwell, 2018), a topic which again I had covered in parts of my *Skills for a Scientific Life* book. With this new book by Bradshaw on the effectiveness of a scientist I felt a real sense of being kindred spirits with him, by which I mean the two of us are scientists from very different science domains facing quite similar, but not identical, challenges. I greatly enjoyed reading his thoughts and advice on becoming as effective as possible. My attention was drawn to the book because of a review of it in the 3–9 May 2018 issue of the *Times Higher Education Supplement* (*THES*) by Dr Jennifer Rohn (2018). Normally in my book reviews I deliberately do not look at other reviews before I read the book and write my own review. The headline statement extracted by *THES* from Rohn's review was that 'This book offers PhDs sound advice but it skirts the improbability of making it.' Rohn also states that

> what fledgling science trainees really need is a good dose of tough love: a comprehensive and honest description of their chances of success, and support and training opportunities to help them understand and realize the rich array of options outside academia. However established scientists assume and preach that doing the right things, along with luck and much hard work will give trainees a fair shot at the professorial pot of gold.

As my *Skills* book had mentioned that in describing my own transition to academic tenure, and *THES* had highlighted my book in their New and Noteworthy section of new books, I wondered if that was Rohn's verdict on my own description. Setting that issue raised by Rohn (2018) aside for the moment, I focus firstly on whether this new book succeeds in its aims. If we look at the Cambridge University Press (CUP) web site for the book it is described as follows:

> What is an effective scientist? One who is successful by quantifiable standards, with many publications, citations, and students supervised? Yes, but there is much more. Truly effective scientists need to have influence beyond academia, usefully applying and marketing their research to non-scientists. This book therefore takes an all-encompassing approach to improving the scientist's career. It begins by focusing on writing and publishing – a scientist's most important weapon in the academic arsenal. Part two covers the numerical and financial aspects of being an effective scientist, and Part three focuses on running a lab effectively. The book concludes by discussing the more entertaining and philosophical aspects of being an effective scientist. Little of this material is taught in university, but developing these skills is vital to maximize the chance of being effective. Written by a scientist for scientists, this practical and entertaining book is a must-read for every early career-scientist, regardless of specialty.
>
> - Written in an engaging and entertaining style, making the topics easy to digest and remember
> - Includes engaging, custom-drawn cartoons illustrating many of the specific topics discussed
> - Discusses sensitive issues, such as personality conflicts and stress management, that are of increasing relevance for the modern scientist, but are usually neglected in academic books.

The book's chapters are as follows:

> *Preface; What is an 'effective' scientist?; Become a great writer; Me time; Writing a scientific paper; Sticky subject of authorship; Where and what to publish; The publishing battle; Reviewing scientific papers; Constructive editing; Fear not the numbers; Keeping track of your data; Money; Running a lab; Making new scientists; Human diversity; Splitting your time; Work–life balance; Managing stress; Give good talk; Getting the most out of conferences; Science for the masses; Dealing with the media; 'Useful' science; Evidence-based advocacy; Trials, tribulations and triumphs; References.*

As I read through the well written prose, enjoying the sketches by René Campbell, I often nodded at advice and shared experience from the vantage point of this very experienced scientist, whose background as a global ecologist gave perspectives often different from mine. The section on giving a good job interview talk to leap that major hurdle from the postdoctoral scientist to a permanent position I thought especially valuable. Thus, notwithstanding Rohn's criticism about this aspect of Bradshaw's book, this is very good advice from Bradshaw. Overall, following the advice given in most of these chapters would yield improvements. There would conversely be little to no harm done by the aspirant not implementing the advice properly. (Mentioning the lovely sketches by René Campbell, I must remark that the book front cover is, however, not for the squeamish.)

When I came to the penultimate chapter, *Evidence-based advocacy*, I was disappointed that like the book on *Scientific Leadership* by Niemantsverdriet & Felderhof (2017) there was no text on the different scientific methods. It was clear that in global ecology the predominant method is hypothesis driven, through the several mentions in *The Effective Scientist* of making a hypothesis, gathering data and then making a statistical evaluation of its significance. This is not asking a question like 'What if the speed of light is finite?' as Einstein did, or saying 'let's make a collection' as Darwin did. The chapter *Evidence-based advocacy* was true to the CUP web site description: 'Truly effective scientists need to have influence beyond academia, usefully applying and marketing their research to non-scientists.' This chapter was, then, to be especially important. It introduced the terminology, new to me, that science is not objective but 'reduced subjectivity'. I can imagine that in global ecology one could well meet some rather hostile critics and advocacy from the scientist being interviewed would be the defensive shield. However, I felt very uncomfortable at this chapter's advice, and in particular I felt that a scientist trying to follow this advice would be in serious danger of harming their science career, not optimizing it. In my *Skills for a Scientific Life* book I had wrestled with the Winston Churchill view that 'scientists should be on tap and not on top', to which I reasoned instead as follows:

> It is vital for us as scientists to be better prepared to face ethical questions and how to do this should surely be a mandatory part of the skills training we receive. One thing is for sure, the discussions of the implications of scientific research discoveries will not be for us to define on our own, such discussions must include all constituents of society, and at the least society's elected representatives. Conversely these elected representatives must include scientists in such debates to provide firm contact with the scientific facts.

Furthermore, to press my point now, in such meetings scientists should be voting members, not just there to answer queries and explain the scientific evidence as known facts, be that the statistical significance of data or the consequences of *e.g.* $E = mc^2$ for peace or war.

So, without the chapter Evidence-based advocacy I would have recommended the book by Bradshaw, but since it is there I am afraid I cannot. To strive to be as fair as possible to the book, Rohn (2018) did not raise any objections about that particular chapter.

I return now to the Rohn (2018) criticism of Bradshaw's book that it skirts the issue of the improbability of a science PhD making the transition to a permanent scientific post. This is indeed a major concern of the funding agencies, and a previous analysis of the biomedical research environment and lack of a career progression of PhDs or postdocs in the USA is very relevant to this (Alberts *et al.*, 2014). The in-depth study and commentary by Alberts *et al.* (2014) documents the undue proliferation of the number of PhDs and postdocs in biomedical research in the USA and then goes on to make specific recommendations for improvements. These include the need for a predictable budget for a funding agency, such as a five year commitment of the USA government (rather than an annual approval method); reducing the number of trainee scientists; increasing the salaries of postdocs; and encouraging the briefing of such scientists about alternative careers. Various of these are already covered in the British funding environment. An additional aspect of the Alberts *et al.* (2014) report is the importance of recognizing that the principal investigator (PI) scientists are not just writing machines, of grant proposals and of publications, but PIs should be encouraged/required to continue their own research and the regular updating of their research skills. Indeed, I firmly encouraged such an approach in my *Skills for a Scientific Life* book, not least as the poor grant proposal success rates obviously commend that one does so or abandon various of one's carefully planned pieces of research. Overall, most importantly, whilst we must present our results as well as possible and, yes, we must strive to win funding with the best written proposals, the heart of the skills for a scientific life are our own practical engagement, or otherwise we become 'solely' a research *manager* rather than a research *scientist*.

REFERENCES

Alberts, B., Kirschner, M. W., Tilghman, S. and Varmus, H. (2014). *Proceedings of the National Academy of Sciences of the United States of America* 111: 5773–5777.

Helliwell, John R. (2016). *Skills for a Scientific Life*. Boca Raton, FL: CRC Press/Taylor and Francis Group.

Helliwell, J. R. (2018). *Journal of Applied Crystallography* 51: 564–566.

Niemantsverdriet, J. W. and Felderhof, J. -K. (2017). *Scientific Leadership*. Berlin, Boston: De Gruyter.

Rohn, J. (2018) "Times higher education supplement", May 3–9 issue.

Appendix A3: Scientific Leadership, *by J. W. (Hans) Niemantsverdriet and Jan-Karel Felderhof*

Reference: John R. Helliwell book review of *Scientific Leadership*, by J. W. (Hans) Niemantsverdriet and Jan-Karel Felderhof (2017) published by De Gruyter. *J. Appl. Cryst.* (2018). **51**, 564–566

This book, *Scientific Leadership*, I gladly accepted to review. I had included the topic in Chapter XVI of my book *Skills for a Scientific Life* (Helliwell, 2016) and so I am obviously interested in it. In her review of my book, Elena Boldyreva (2017) drew my attention to the book by Hans Selye (1977), a scientific leader of the time in physiological aspects of stress, which I immediately purchased. A book by Kathy Barker (2010) focuses solely on leading one's laboratory. I had devised my *Skills* book to include aspects of the scientific method and various roles I had played, but it is also a manual comprising many chapters, each being a 'how to' description which readers can dip into as needed. I included the harnessing of the tools of managerialism but also considered its ills, at least as seen in universities by myself and often described in the *Times Higher Educational Supplement.*

This book by Niemantsverdriet and Felderhof focuses on scientific leadership and is geared towards a person becoming a leader and to following managerialist techniques. As I read the contents pages I wondered about the last chapter's *How to implement self leadership and the 3B-6T-9E enhancement philosophy in your organization.* So I began …

The contents of the book are as follows:

> *Foreword; Acknowledgments; Special foreword; Contents; 1. Creating success in your own scientific world; 2. Leading yourself and others in research; 3. Presenting science: talks, publications, posters, and some ideas on conferences; 4. Management skills for researchers; 5. Planning the road to your future in science: strategy and comprehensive plans; 6. Understanding how to build successful teams: creating synergy between people who trust each other; 7. On the road to scientific (self) leadership; 8. Post scriptum – implementation; Worksheets; Index.*

The front cover is labelled 'GRADUATE'. The back cover states the book's aims: 'Modern Science is teamwork. But how can young academics go from being a productive member of a scientific team to leading their own? Entry level positions for PhDs in Science often require the infamous 'people skills'. The authors aim to equip young academics with the right ideas and strategies for their scientific leadership development. Become a successful leader not with tricks, but with an inspiring and straightforward vision and mission, the correct mindset, and effective teamwork.'

On page 9 I learnt why the book cover image was chosen: '*... people have to understand the goals and needs of their organization thoroughly, and they need to dispose over sufficient freedom to operate and decide. Terms like 'self-steering', 'self-management', and even authentic 'self-leadership' have entered the scene; hierarchical structures tend to become less dominant. In its ultimate form, third-generation governance and control rely on unique abilities and strong points of individuals, who together operate like a flock (see cover illustration). The situation determines which abilities are needed most, and as in an effective flock of geese, the leadership changes according in an almost automatic fashion.*' I was unsure at this point what the message was. Personally I think it important to assess any theories of what scientific leadership is with some very well known examples. For such I always start with Albert Einstein. We can immediately see that his huge leadership example does not fit a 'flock of geese' model. However, as Abraham Pais's superb biography of Einstein shows (Pais, 1982), Einstein did refrain from citing others when he really should have done, notably Lorentz and Poincaré in his 1905 special relativity paper. So he might be deemed in a flock of three geese, to be fair.

From the book's section on presentations a reader will get some clear and very good advice, such as page 60 – '*The best advice on making effective slides is probably to remove all information that is not strictly necessary.*' – although I am unsure why they qualify it with 'probably'. We are also shown an excellent example of the wrong and right way to present a rather complicated set of curves so that a reader can '*immediately concentrate on the meaning of the data*'. There are some oddities in the authors' coverage, however. They seem to assume a standard length of a talk being 20–30 min. They also advocate the use of intermediate conclusions being mentioned at the end of each section of a talk. They do not mention the main hazard of this, which is the audience assuming an earlier finish than the speaker has planned. At the IUCr Hyderabad Congress in the commercial exhibition I saw on the Springer recent and new books display that there was a whole book on scientific presentations (Alley, 2013). I bought a copy and read it cover to cover. I was impressed with Alley's advice and immediately set about adding some further polish to one of my recent presentations, which had already gone very well. This illustrates the depth and breadth to which one can go on these topics.

Chapter 4 of Niemantsverdriet and Felderhof's book covers a variety of management skills for researchers. This I found to be generally good and sound. But again there were odd things about it. Firstly, the tool of risk estimation and mitigation seems to me obligatory and incredibly useful but was not covered. Secondly, the example on page 86 involved a project to be managed on a Dean's instructions for developing a teaching course in computer skills for first year students, i.e. not a research project example at all. Thirdly, the traffic lights system of managing projects, namely red to stop a project, amber to pause it and green to continue it, is simple but effective, I find. This, too, is not discussed. Fourthly, for a scientific leader to aspire to executing a project so well that its Gantt chart might be '*a nice wall chart in your laboratory to show visitors*' I found rather strange. If what we are about is science and discovery, then much hoped for is the unexpected finding, be it graphene or a pulsar to name two of the best-known unexpected discoveries, which of course will be outside a Gantt chart's planning and scope. The strengths of the authors in approaching people and situations with managerial tools come to the fore with the final section of chapter 4, '*Conflict management*'. This section concerns conflicts between team members in a laboratory and the strategies that can be adopted to move on. The authors link the different ways of dealing with such conflicts with their earlier chapter 2 (e.g. Fig. 2.1) on the '*four operational personality dimensions*' and types of personalities. They also have a very nice section on *Relationship management*, namely the developing of good contacts and retaining them. A further example of harnessing good tools that stem from a managerialist approach is their advice on the systematization of consulting one's students, colleagues or coworkers ('the team'). This includes brainstorming together and use of green, yellow and red coloured paper notes for identifying success, issues needing attention or failure, respectively, in one's organization. This can then also be mapped onto a diagram of Vision, Strategy and Outside relations as shown in their vivid Fig. 5.7 on page 106. On page 110 we learn the meaning of the '3B Principle': '*be oneself, belong to the group and be valued*'. This notion is expanded on in Section 6.4 entitled *The 3B approach to authentic resources*. The authors see them as '*essential prerequisites for establishing trust, mutual respect, an open culture for critical and constructive exchange of ideas, and pride in achieving shared goals*'. I feel that the language leaves scope for confusion here and I find it difficult to know what the authors really intend. That said, the authors do bring out in a sparkling way one of their other tools, the understanding of different personality types in a research group, which allows one to see what happens, or might happen, in teams with differing examples of mixtures of personality types (pages 115–117). On page 125 we learn the meaning of 3B-6T-9E, and I quote:

> *Forgive us for dubbing our approach to (self) leadership as the 3B-6T-9E concept.*

3B (be oneself, belong to the group and be valued) the three requirements for authentic behaviour

6T as the six stages in the trajectory to realizing a grand plan (Triggers, Talents, Thrives, Thrills, Trails and Track)

9E as the nine elements for developing skills and building up experience (including skills for autonomous and team performance and devising strategies for creating high impact).

This is too much jargon, I think.

On page 134, Fig. 7.4, we learn the core of why they value the team *versus* the individual. But the approaches of both an individual and a team are needed, I think, and great things can be achieved by both. So, in all this managerialism, I felt, there is too little acknowledgement of what science is, be it the variety of scientific methods (Helliwell, 2016) or the very philosophical aspects such as articulated by Chalmers (1999) or Selye (1977). But common sense and good analysis come to the fore on page 143 with the authors' analysis of encouraging and discouraging workplace environments. A major strength of the book is its guest columns. Advice from successful leaders is invaluable, and the authors have chosen well from ones they know personally; on pages 14–15 Jens Rsitruo-Nielsen, a former research director in industry and a founding member of the European Research Council, writes about the concept that *Research should be managed by motivation not by control*; on pages 71–72 Prof Graham Hutchings, Professor of Physical Chemistry and Director of the Cardiff Catalysis Institute at Cardiff University, writes on *Time, money and great ideas*; and on page 124 Yong-Wang Li presents a motivational vision on how *Science is the right guide for a safe and sustainable future for our world.* I have highlighted aspects above, the good and the questionable, in an evidence-based way. What do I think is missing from the book? A gap in coverage is what to do (*i.e.* practical advice) about bad leadership. The authors may quite reasonably say such leaders should read their book and improve themselves! There also could be some comment and guidance for scientific leaders on achieving a proper gender balance. This book would benefit from having a much bigger ethical dimension, I think. Overall would I recommend this book? Actually I would because its intent is very sincere, although the language and style I identify with managerialism, which I also rebel against, creative spirit that I am. But that said I can see good tools from managerialism that can be of help in realizing one's scientific dreams, ideas and hunches (my terminology).

REFERENCES

Alley, Michael (2013). *The Craft of Scientific Presentations. Critical Steps to Succeed and Critical Errors to Avoid.* New York: Springer Science and Business Media .

Barker, Kathy (2010). *At the Helm. Leading Your Laboratory.* New York: Cold Spring Harbour Press.
Boldyreva, E.V. (2017). *Journal of Applied Crystallography* 50: 1241–1242.
Chalmers, Alan F. (1999). *What is This Thing Called Science?* 3rd edition. Buckingham: Open University Press.
Helliwell, John R. (2016). *Skills for a Scientific Life.* Boca Raton, FL: CRC Press.
Pais, Abraham (1982). *Subtle is the Lord: The Science and Life of Albert Einstein.* Oxford: Oxford University Press.
Selye, Hans (1977). *From Dream to Discovery.* On Being a Scientist. New York: Arno Press.

Appendix A4: Managing Science: Developing Your Research, Leadership and Management Skills, *by K. Peach*

Reference: John R. Helliwell book review of *Managing Science: Developing Your Research, Leadership and Management Skills* by Ken Peach (2017) published by Oxford University Press, Oxford. *J. Appl. Cryst.* (2018). **51**, 1773–1776.

The title *Managing Science* immediately raises several questions: Who in universities, research institutes or large-equipment installations would benefit from such a text? The scientist who is going to take on some management task because they wish to get promoted? The manager who does little or no science? The scientists who have to endure an apparent complexity of management over their heads and seek illumination as to just what is going on? The sub-title *Developing Your Research, Leadership and Management Skills* suggests it is for the heads of department alone. If so, will the book be solely retrospective? Or will it also offer prospective solutions, for example addressing the agreed ills of the science workplace the world over, such as male-dominated workplaces? I should mention that I worked alongside Ken Peach when he and I were directors of different aspects of the Council of the Central Laboratories of the Research Councils, the CCLRC. I observe that Ken Peach is very well qualified to address the topic of this book, with more than 40 years of experience managing science, in the UK and at CERN in Geneva, including 25 years at a UK university, as he puts it, managing his own science. As for the book's aim, the back cover states it is 'to introduce the working research scientist to the art and techniques of management and the skills necessary to be a good and effective manager and leader of science and scientists'. The Preface adds 'Balancing the needs of science – which has to be free to follow the trail, wherever it may lead – with the demands of society – that it be accountable, responsible and, if possible, useful – is a skill that needs to be carefully developed.' Let us see what the answers to the various questions posed above are.

After Chapter 1, an *Introduction* of two and a quarter pages, Chapter 2 defines and sets the boundaries of the fundamentals, namely *Science,*

Research, Development and Scholarship. Of these four terms and their domains, scholarship is the most difficult to define but this is skilfully done. Chapter 3 continues the domains, this time in the setting of *Universities and Laboratories.* With a book of this sort, knowing quite where to pitch is difficult. The author opts for a down to the very basics approach and describes a university as an organization. On page 14 he states that 'The management of undergraduate teaching is beyond the remit of this book, as is some postgraduate education, namely that which leads to a taught diploma or a master's degree.' I confess to some disappointment in that decision: I consider an ideal university to be one that intertwines teaching and research such that in an undergraduate course an academic sets individual and team projects at or near to the research frontiers so as to be both not like undergraduate laboratory experiments and yet achievable in their goals. Such projects are also, besides a learning exercise, ones where both incremental and innovative research can be attempted by the first-degree student. The layout of Chapter 3 seems to me rather odd. It has section 3.1 on *Universities* and, later, section 3.4 on *Universities as businesses.* One issue that Chapter 3 might have wrestled with is the rise of managerialism in universities, adopted from the large laboratories or what might be bundled together as the 'scientific civil service', admittedly a UK terminology. Since the book is about managing science, and thereby de facto its utility, the alternative, cynical, view of management, i.e. managerialism, is outside its domain.

Chapter 4 is the first to get down to business in describing *Leadership, Management and Communication.* In section 4.1 I found the opening unpromising: 'There are leadership opportunities and challenges at all levels – from the leader of a small section of two or three people … to the organization of several hundred scientists, engineers, technicians, administrators and support staff.' By unpromising I mean that this of course presumes a limited definition of the scientific method, since an Einstein working alone is not included. But, to be fair, managing oneself involves a different terminology, such as one's skills. The author of this book is highly experienced as a director and makes the careful, and very important, distinction about whether an organization wants a director to provide leadership or to administer it. The distinction is that an individual leading from the front, as I would call it, sounds good to the individual but could be seen as a loose cannon to the organization. This section of the book shows a depth of experience, although it avoids the psychological personality profiling formulations that might be usefully harnessed in running a team (Niemantsverdriet & Felderhof, 2017). Section 4.2 on *Management* offers generic tips, including the interesting pros and cons of when a director has their door open or closed. Section 4.3 on *Communication* has equally helpful tips. The final section, *On appointment*, is aimed obviously at the new managing science recruit. This seems almost like a manual of 'how to be a politician': good words to be used on facing a first group, *i.e.* staff, meeting being 'evolve', 'look for new opportunities' or 'building upon' as 'these signal change without being specific or scary'. The previous section on communication offered similar 'political' guidance with its terminology: 'paltering – not exactly lying but equally not telling the truth'. In summary, 'organizations can survive a leader who cannot manage or communicate but who can delegate but they cannot survive with a leader who cannot lead'. This, however, contradicts the careful distinction made above, that leading from the front may be viewed as being a loose cannon by the organization.

Chapter 5 is *Building Research Teams*. This comprises sections of the book logically proceeding from graduate students to postdocs, fellows, engineers and technicians, administrators, and visitors. The latter, *Visitors*, is a section of considerable breadth and hints at serious problems, which fortunately occur in very few cases. Section 5.7, *Team spirit*, I think is very good but does need a section on, in effect, dealing with difficult people. Section 5.8, *Large projects*, gets into top gear, as one might expect for the author's big science background in particle physics.

Chapter 6, *Recruitment*, is written with a great depth of experience obviously, as becomes very clear when the author mentions near the end of the chapter that he has written hundreds of letters of reference and read thousands of them as an interviewer. The chapter is very thoroughly done. In Equal opportunities, section 6.5, it seemed odd to me that the UK's Athena SWAN (Science Women's Academic Network) gender equality scheme, adopted now in other countries, was not mentioned, and its merit awards for equal opportunities' achievement of science departments on a hierarchical scale of bronze, silver and gold. Table 6.2 has a helpful list of forbidden questions, *i.e.* that cannot be asked of candidates at interview.

Chapter 7 is *Managing Scientists and Others*. This chapter brings us very close to the title of the whole book, clearly then a very important chapter for its aims. A manager of science is going to need to measure performance and this is where the chapter starts, with section 7.1. The chapter moves nimbly from the manager to the managed and their respective participations in the process, which should be SMART (specific, measurable, achievable, realistic and time limited). These innovations from the scientific civil service into university life have improved science in academe I think, as I explained in my own book *Skills for a Scientific Life* (Helliwell, 2016). Section 7.5 on dealing with problems starts with managing poor performance and deals with what is surely one of the most important aspects of any manager's role. The advice given I found clear and to the point and with sufficient detail of common types of case to be useful to the manager of science project teams. The section that is missing from this chapter is *Managing your boss*.

Chapter 8 is on *Cooperation and Competition*. After a general introduction, section 8.1 covers the topic in academia. The summing up paragraph which is highly formal may be how it works in particle physics. Rather, in my areas of science, one proceeds by deciding if you might be able to work with a person whose expertise complements your own and your laboratory's emphases, then building up from small to larger projects. Section 8.2 is on collaborations with industry and is nice and clear on the necessary formalities and full of experienced insights in several very good case studies. Section 8.3 on *Multidisciplinary research* is equally on the mark.

Chapter 9, *Councils, Boards, Committees and Panels*, is a thorough and clear treatment of the topics.

Chapter 10, *Committee Meetings*, again is a very thorough treatment of the topic. In my experience committees can be incredibly effective but, as the author explains well, need to be properly constituted and run. The last section, Implementing cuts in budgets, immediately raises the question of whether in fact a committee of stakeholders is the management tool to implement them.

Chapter 11, *Reviewing Research, Making Proposals and Evaluating Science*, shows the author in full stride with his wealth of experience of these. A thorough description of various types and situations involving peer review, and writing one's own proposals anticipating the peer review to come, is given. The examples and anecdotes keep the chapter alive, along with its naturally fairly mundane but essential details. An unexpected gap is the absence of a description of the peer review of articles submitted for publication and, increasingly, the need to scrutinize the underpinning data to improve the reproducibility and quality standards of research narratives in journals. In the large-scale facilities section, practical criteria could have been offered for judging whether a facility is operating at a world-ranked level. There are well established metrics at such facilities, whether in Europe, the Americas or Asia, for this and peer review, the subject of this chapter, extends to facilities. This is an important aspect of managing science well, and, being a big team effort for facility staff and many thousands of principal investigators, is where the military style works at its best, i.e. managerial methodology is essential here. The section on *Preparing for an institutional review* is a gem of clarity and the questions posed for self evaluation by the institute in preparation for a review could be equally well applied to an individual academic scientist and their laboratory aims, however small or large their laboratory is. Section 11.5 is a long section. Its bulk is a masterly description, and dissection, of the UK's efforts with its Research Assessment Exercise, and its descendants. Similar exercises have been undertaken in other countries. The most serious flaw of such RAEs is how then to reach objectivity. This is addressed by the author of this book but he follows the dictum that judgement is solely what is needed. Several times he wholesale rejects the objective evidence of the various research assessment metrics on offer, which I would say can and should be moulded into the judging process.

Chapter 12, *Managing Projects*, is a thorough description and will be very useful to the reader persevering to the end. But some project failures would have both been instructive and added some spice. Famous ones that come to mind are the closure of the Supercollider's construction in Texas and the mis-polished mirrors of the Hubble Space Telescope. For balance, some praiseworthy ones would be good, e.g. the wiggler 5T cryomagnet for the Synchrotron Radiation Source (SRS) at Daresbury, which was built by the Rutherford Appleton Laboratory; this was a ground-breaking device in synchrotron radiation production, with pioneering applications in various scientific fields.

Chapter 13 on *Risk* is a vital topic in planning and managing science everywhere, as I have seen serving on numerous international advisory committees. Again the author gives a masterly description in the sections on *Evaluating and managing risk*, on *The risk register*, and on *Risk, reward, investment and progress*. Tables 13.1 and 13.2 are very helpful, as are Figures 13.1 and 13.2. Section 13.3 on *Reputation risk*, although lucid, important and clear, does not sit well in this chapter and would fit better I think in a dedicated ethics chapter.

Chapter 14 is on *Health and Safety*, a key part of managing science. This description is generally good. Some mention could have been made of chemical hazards and their safe-as possible use, the 'fail safe design' of equipment, radiation protection, and safe use of lasers, to mention a few of the more common areas of scientific research requiring 'safety rules'. But obviously a dedicated description of each would not really be feasible in a book of this type.

Chapter 15 is on *Dealing with Disaster*. A nice explanation is given with the practical example of magnets failure at CERN in 2008, which delayed the Large Hadron Collider (LHC) project for a year. I recalled various aspects of this from memory of the media coverage at the time, and so the chapter's principles and practice related well to business continuity planning of the LHC research programme example. Also, the examples of data recovery planning were well chosen. Even as individuals, we also know the virtues of regular disk backups, and I would emphasize that as advice to my students. The final section on insurance I did not particularly like and disagreed with the life insurance analysis analogy offered; if you are a major wage earner for the family, you have to provide for your dependants; it is not a 'saving money in a jar' type of thing.

Chapter 16 is on *Problems*. Page 226 provides a pretty comprehensive list of what is to be covered in the chapter and indeed what a manager might face. Very good generic examples follow of poor performance examples: due to alcohol abuse or misuse of computers of various kinds, and a remuneration grievance example. The category of bullying is not dealt with in detail although it is commonplace, not least in institutions under the great pressure of research and teaching performance assessment, where managers can seem to 'simply' transfer these pressures to their staff. The obvious conflict of interest here for the manager is a major complication in fair handling of staff complaints against department heads. So, the role of the trade union can be vitally supportive in ways that the employer and their managers in effect are not. Staff exiting from their employment can be confidentially canvassed to track such problems at an employer; indeed, this is a tool used to check for diversity discrimination, such as a department with a bullying culture. This is identified as being a significant contributor to male–female gender disparities in staffing. The chapter concludes with details of preparing for difficult interviews. The author is to be commended on being candid and sharing his approach; the context I felt was not good, however, and presumably represents the approach of a particular employer and their managerial training sessions. An emphasis on having prepared negotiating positions and 'sides' clearly suggests a wish for an outcome on behalf of the organization. An unsatisfied employment staff grievance can escalate into legal actions such as an employment tribunal process. This is then a big topic only lightly touched on, with variations of legal process in different countries.

Chapter 17 is somewhat mysteriously entitled *Just Managing*, but the author neatly, and you will just have to buy the book to find out, explains its threefold pun. Section 17.1 is on *Ethical management.* Here we touch the nerve exposed already in the previous chapter. Interestingly the principles quoted are from the UK Engineering Council and are good, as far as they go. But science and engineering are not one and the same; what does not appear in the list is authorship or data or peer review in science. These are, however, dealt with in other parts of the book, which is a fair defence of it. Section 17.2 is *The stressed manager.* The text here is from a highly experienced manager and I would largely concur with it, but the 'schedule meetings starting at 4 p.m.' advice could be seen as one of the causes of the male-dominated workplace, especially in the higher positions of an organization. The time management tips although insightful miss the very useful one I learnt on a course, namely the time management quadrant and how one should personally strive to be in the important not urgent quadrant. Also, any situation forcing you into the important and urgent quadrant, *i.e.* firefighting, should be treated in two steps: put the fire out and then make sure it does not happen again! Section 17.3 is *Management is management.* It contains sage advice on accounting, auditing, grant management, and laboratory and office space management.

Chapter 18, *Summary and Conclusions*, is just that, but also offers a valedictory speech. It makes clear that 'Although I enjoyed my years in the various senior management positions, I cannot say I enjoyed them more than I enjoyed doing science.'

So, can one 'herd cats', where the 'cats' I define here are creative, free thinking, imaginative scientists? This book convinces me that they can be, at least when united by the vision to achieve a grand science challenge or have a multi science large-equipment user facility. It is indeed the classic image of particle physics united to find, say, the Higgs boson. However, could an old-days Einstein or a modern-days Higgs have been 'managed'? Does one even need a person in control who is a scientist, or does having a non-scientist in charge result in objective leadership?

Overall this book is an excellent manual and assistant to the professional scientist. It is carefully prepared by a highly experienced senior scientist. There are some caveats, which I have indicated specifically in this book review; these could be readily addressed in a second edition.

Acknowledgements

I am grateful to the Book Reviews Editor Massimo Nespolo for his comments on this book review which improved it, notably capturing better the international aspects.

REFERENCES

Helliwell, John R. (2016). *Skills for a Scientific Life.* Boca Raton,FL: CRC Press, Taylor and Francis Group.

Niemantsverdriet, J.W. (Hans) and Felderhof, Jan-Karel (2017). *Scientific Leadership.* Berlin, Boston: De Gruyter.

Appendix A5: Writing Chemistry Patents and Intellectual Property: a Practical Guide, *by Francis J. Waller*

Reference: John R. Helliwell book review of *Writing Chemistry Patents and Intellectual Property: A Practical Guide*, by Francis J. Waller (2011), published by John Wiley & Sons, Hoboken, NJ. *Crystallography Reviews* (2017). **24**, 131–132.

This book describes itself as follows:

> Intellectual property is constantly at risk, and the protection of chemical science and technology through the patenting process allows individuals and companies to protect their hard work. But in order to truly be able to protect your ideas, you need to understand the basics of patenting for yourself. A practical handbook designed to empower inventors like you to write your own patent application drafts in conjunction with an attorney, Writing Chemistry Patents and Intellectual Property: A Practical Guide presents a brand new methodology for success. Based on a short course author Francis J. Waller gives for the American Chemical Society, the book teaches you how to structure a literature search, to educate the patent examiner on your work, to prepare an application that can be easily duplicated, and to understand what goes on behind the scenes during the patent examiner's rejection process. Providing essential insights, invaluable strategies, and applicable, real-world examples designed to maximize the chances that a patent will be accepted by the United States Patent and Trademark Office, Writing Chemistry Patents and Intellectual Property is the book you need if you want to keep your work protected.

There are 15 chapters, then a section with the answers to questions posed at the end of several chapters and finally, also very helpful, several example patents quoted verbatim.

At the Amazon page for this book there are several customer reviews including from patent agents and attorneys. C. Richard writes (March 2012)

> I think that overall, the author did a good job in structuring the book. There is much useful information in it, and it seems well presented for the intended audience, at least overall. The book focuses on a chemical technology context, but at least much of it could be applied in other technology contexts.

M. Cox writes (January 2012)

> It is not a replacement for a good patent attorney. Reading this book in no way qualifies someone to write their own chemistry patent.

T. Forge (April 2013) writes

> Many parts of this book are now out dated due to the enactment of the American Invents Act … Disclosure – I am both a technical expert (not in chemistry) and a patent attorney. This book obviously wasn't geared toward me. I still found it useful because I often talk and advise people regarding their desire to apply for a patent or what is going on with their application.

I found it interesting that these people, expert in the area, took the time and trouble to lodge their views.

As an academic research scientist I found this book very interesting. I have had several instances in my career where I consulted my University of Manchester Intellectual Property (IP) office. In one case it went so far, with their help, to be issued with a one year provisional UK patent on a novel apparatus for recording polychromatic X-ray diffraction patterns from a crystal, i.e. in a 3-D arrangement. It was not taken further because the likely market was in the end deemed too narrow and thereby not worth the cost of a full patent for 20 years and worldwide. With the increasing business orientation of universities, it will be more and more the case that academics will need to be careful to consult their IP office before releasing any results. Furthermore Waller's book, or others like it, will be more and more a part of the academic's training. In the end, as Waller's book emphasizes, the academic is very unlikely to engage in the writing of a patent application on their own. Nevertheless we must be better informed than I was through my early career. Waller's book does just that and effectively I think. It is clear that his experience in industry and in the training courses he gave as part of the American Chemical Society effort in continual professional development really shine through.

In the book there are some great insights made by Waller. At page 97, 4 to 6 lines up: 'Scientists ask questions and then find answers by performing creative research. On the other hand, inventors look for problems and then find solutions by combining their creative ability with research.' At page 52 ' … an invention must be a new and useful process, machine, manufactured article or composition of matter or any new and useful improvement to any of those four'. At page 175 'One way to overcome the obviousness rejection by a patent examiner is to be sure your invention gives new and unexpected results not predicted by the prior art and that the results are superior to previous inventions.'

For good measure the book also has a chapter on copyright (more familiar territory for an academic) and trademarks. There is also a very helpful chapter on the various educational resources on intellectual property where the reader can go and learn more. This book should definitely be read by a modern academic. The book has quite a number of typographical or syntax errors, almost as if the proofs were not read properly by the author or publisher. However, none interrupted my reading very much and did not create any misunderstanding for me. As the reviewer at Amazon remarked, quoted above, patent law apparently does change but realistically I think any academic would consult their University's IP office for assistance in a real case. This book for me is more about becoming much wiser about the whole process and even the fundamentals of how we approach our curiosity driven science or the technical challenges that do arise.

The book has one prevailing assumption: page 142

> Without the patent or the exclusive right associated with a patent, an individual or company would not have sufficient incentive and time to bring their new product to the marketplace. (Furthermore) If you are going to manufacture and sell your product in several different countries, then you will need a patent in each.

Quite so, but I do always recall the Jonas Salk story regarding his discovery and development of one of the first polio vaccines: When asked who owned the patent to it, Salk said, 'Well, the people I would say. There is no patent. Could you patent the sun?' Quite so.

Appendix A6: The Scientific Method: Reflections from a Practitioner, *by M. di Ventra*

Reference: John R. Helliwell book review of *The Scientific Method: Reflections from a Practitioner* by Massimiliano di Ventra (2018) published by Oxford University Press, Oxford. *J. Appl. Cryst.* (2018). **51**, 1509–1510.

The contents of this short book involve 18 chapters and 128 pages. The chapter headings are 1. *Science Without Philosophy,* 2. *Material World and Objective Reality,* 3. *First Principles and Logic,* 4. *Natural Phenomena and the Primacy of Experiment,* 5. *Observation and Experimentation,* 6. *The Role of Human Faith in Science,* 7. *Approximate and Limited Description of Natural Phenomena,* 8. *Hypothesis,* 9. *Theory,* 10. *Competing Theories,* 11. *Can One Theory Be 'Derived' from Another?,* 12. *Verifying or Falsifying? And What?,* 13. *Don't Be a Masochist!,* 14. *'Consensus' in Science? What Is That?,* 15. *Flow Chart of the Scientific Method,* 16. *The 'What' and 'Why' Questions,* 17. *'Scientism': Abusing the Scientific Method,* 18. *Final Thoughts.*

The back page of the book describes it as follows:

> This book looks at how science investigates the natural world around us. It is an examination of the scientific method, the foundation of science, and basis on which our scientific knowledge is built on. Written in a clear, concise, and colloquial style, the book addresses all concepts pertaining to the scientific method. It includes discussions on objective reality, hypotheses and theory, and the fundamental and inalienable role of experimental evidence in scientific knowledge.

From the author's university web page I learnt that he is a theoretical physicist and has a strong publication record: for ease and simplicity I quote his h-index of 63. The author concludes his Preface with 'My hope is that this book will rekindle a much needed interest in the centuries-old structure and foundations of this marvellous human enterprise we call Science.' Has the fire gone out? Let us find out …

Chapter 1 sets the scene by describing that science and philosophy are different. This concept is most neatly summed up on page 13 at the end of chapter 2: 'All scientific statements need to be logical, but not all logical statements need to be scientific.' Chapter 2 also sets out boldly and clearly that there is an objective reality: 'Denying the existence of an objective reality that is present irrespective of the observer is equivalent to denying the very existence of Science and its method.' This sentence is preceded by the author's statement that 'I have encountered people that would deny the existence of such an objective reality. Although this is often the subject of movies or some 'beliefs', I cannot truly fathom a scientist's claiming the same.' I would definitely agree. Chapter 3 is entitled *First Principles and Logic* and 'summarizes the core of what is called common sense'. It is made clear by di Ventra, for example, that 'For any proposition, either that proposition is true or its negation is true' and 'For an event to occur at a given time, there must be another event that caused it.' Then the chapter usefully makes clear that a scientific description 'needs to lead to testable predictions'. Chapter 4, *Natural Phenomena and the Primacy of Experiment*, opens with a cartoon sketch of the leaning tower of Pisa linking, I imagine, to Galileo's experiment of the late 16th century that two objects of different masses would both fall to the ground at the same time as a result of their experiencing the same acceleration due to gravity. That picture is somewhat undermined when on page 34 in chapter 5 the author outlines the situation where two objects of the same mass fall to the ground at the same time, excluding air resistance, and so seems to miss Galileo's point. Chapter 5, however, entitled *Observation and Experimentation*, actually seeks to stress that 'controlled laboratory experiments on a particular phenomenon may not always be easy to perform'; hence the example of a feather and a ball falling to the ground is a fair choice. Chapter 6 is entitled *The Role of Human Faith in Science*. The author advances that we 'trust the textbooks and journal articles that describe *e.g.* how many planets there are in our solar system and *e.g.* that atoms support discrete energy levels and so on.' Actually, I would rather emphasize that school laboratory classes in physics, chemistry and biology are essential for students and pupils. Thus they could directly see in their school laboratory key scientific evidence. These direct experiences would ideally include seeing *via* the eyepiece of a prism spectrometer the discrete spectra of sodium gas emission and use of their school's telescope to view our solar system planets. So, I would argue that the faith of the public should not come into it, indeed that matters of faith would be a concern of religion not science. The strength of science is that one does not have to take someone else's word for it. Chapter 7 is entitled *Approximate and Limited Description of Natural Phenomena*. This chapter is rather vague in its mission. It seems to be making the obvious point that mankind has not discovered everything as yet (quote: 'Takeaway message: The description of phenomena is always incomplete.'). Chapter 8 is entitled *Hypothesis.* This would be a cornerstone chapter. Four pages into it, however, the power of hypothesis making in science and discovery is still not articulated; rather it is stated that 'Hypotheses are not objective data or facts.' No, but I would suggest that they are a marvellous instrument for planning new experiments. Chapters 9, 10 and 11 focus on theory. Chapter 10 is promisingly entitled *Competing Theories.* This chapter has an interesting start: 'Suppose I come up with two distinct theories for a given set of observations ... Which one should we choose?' An instructive example that the author could usefully have described would be the two theories of light: light as waves and as photons. An interesting excursion is made by the author into the role of money, funding and clout, *i.e.* reputation in modern science. The final sentence takeaway message captures the essence: 'If the theories make different predictions, then experiments should settle the issue.' It would have been good to have an important practical example here, for example, is there climate change or not? Chapter 11 is entitled *Can One Theory Be 'Derived' from Another?* This I found to be a very sound chapter. A telling example was comparing Newtonian mechanics with Einstein's theory of special relativity. The latter has a vital difference: the speed of light is taken as a constant by Einstein, and what does that lead to in consequence? One result was the abandonment of the need for an ether as a medium for light to propagate through, something that fascinated late 19th century physicists. Chapter 12 is entitled *Verifying or Falsifying? And What?* Clearly this chapter is going to be classic Karl Popper axiom driven. The chapter starts with a very amusing cartoon of two scientists debating the theory of everything: 'does this achievement put us out of work?' Unfortunately, the author does not develop the notion of falsifiability of a theory being more powerful than attempting verifiability of a theory. Chapter 14 is *'Consensus' in Science? What Is That?* This chapter provides the chance for the author to introduce Kuhn's paradigm shift as a concept for marking scientific progress. Di Ventra again makes an interesting link with the need to win funding and the fact that for young scientists to get promotion they are compelled to remain within 'the establishment'. In connection with this, I have argued that since funding success rates are consistently never better than one in three, and often worse, it is vital to keep one's personal research, *i.e.* at-the-bench skills, up to date. That way one can still tackle unfunded ideas and/or the adventurous ones that are so hot one would not delay them with writing of a grant application. Chapter 15 is a *Flow Chart of the Scientific Method.* The small-format size of the book makes this figure illegible without a magnifying glass, but with it I think its contents look fine. OUP could have done better with the sizing of this figure. Chapter 16 is entitled *The 'What' and 'Why' Questions.* This is a good chapter as it describes examples of questions which cannot be addressed by science but scientists may well be asked about. Examples given are 'What is gravity?' and 'Why is there gravity?' This leads on to questions posed by the author about the existence of God or intelligent design, thus defining a boundary between science and religion.

A popular science news item in recent years has been the Higgs boson as the God particle, *i.e.* explaining why matter manifests at all, or as Lederman & Teresi put it *The God Particle: If the Universe Is the Answer What Is the Question?* (Dell Publishing, New York, 1993). This could have been a more telling example and certainly one that is raised with me at my tennis club. Chapter 17 is entitled *Scientism: Abusing the Scientific Method.* This is a somewhat convoluted chapter. Suffice to say it is making the point that there is a difference between science and religion. Another way of making the point raised, and the one usually adopted by scientists, is that religion involves matters of faith whereas science is objective, which the author has stressed earlier in his book. Chapter 18 is *Final Thoughts.* This chapter concludes on an excellent note: 'if the scientific method is applied with intellectual honesty, it will allow us to make sense of the facts that Nature will reveal in the process of discovery … objective reality is there.'

In summary I found this book good in parts. Such a short book, with its clear takeaway messages, is needed. A general question I would pose: is there one scientific method?

Certainly the making of a hypothesis is one, stressed by this book. There are others. There is empirical gathering of facts from which a law (in physics) may be deduced. In biology, what about the making of a collection as Darwin did or, in its modern guise, sequencing all the genes of a genome with no initial hypothesis in mind? After the collection is made, progress can be achieved by intellectual reflection. There is also the case where we ask what if we do this or that in our experiments, *i.e.* following a hunch, not as firm as a hypothesis.

Then there is asking a question such as Einstein's 'What if the speed of light is finite?', from which a mathematical theory flows. In a further approach, that involving statistics, CERN, for example, insisted on a '5 sigma' level of proof for the Higgs boson. Another feature of this book is the writing style. The back cover states that the book is 'written in a clear, concise, and colloquial style'. Should colloquialisms in fact be avoided in presenting to audiences from around the globe so as to avoid misunderstandings and be clear? That said, each chapter usefully concludes with a list of take home messages.

Bibliography

Chalmers, Alan (1999) *What Is This Thing Called Science?* 3rd edition. Buckingham: Open University Press.

Greenfield, Susan and Singh, Simon, Editors (2003) *The Science Book 250 Milestones in the History of Science*. London: Wiedenfeld and Nicholson.

Mayr, Ernst (2007) *What Makes Biology Unique?* Cambridge: Cambridge University Press.

Perutz, Max (1991) *Is Science Necessary? Essays on Science and Scientists*. Oxford: Oxford University Press.

Bibliography

Index

Printed in the United States
by Baker & Taylor Publisher Services